Raymond Williams's Sociology of Culture

Raymond Williams's Sociology of Culture

A Critical Reconstruction

Paul Jones
School of Sociology, University of New South Wales
Australia

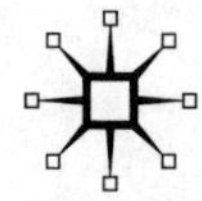

First published in hardback 2004
First published in paperback 2006 by
PALGRAVE MACMILLAN
Houndmills, Basingstoke, Hampshire RG21 6XS and
175 Fifth Avenue, New York, N.Y. 10010
Companies and representatives throughout the world

PALGRAVE MACMILLAN is the global academic imprint of the Palgrave Macmillan division of St. Martin's Press, LLC and of Palgrave Macmillan Ltd. Macmillan® is a registered trademark in the United States, United Kingdom and other countries. Palgrave is a registered trademark in the European Union and other countries.

ISBN-13: 978–0–333–66662–3 hardback
ISBN-10: 0–333–66662–3 hardback
ISBN-13: 978–0–230–00670–6 paperback
ISBN-10: 0–230–00670–1 paperback

This book is printed on paper suitable for recycling and made from fully managed and sustained forest sources.

A catalogue record for this book is available from the British Library.

Library of Congress Cataloging-in-Publication Data
Jones, Paul
Raymond Williams's Sociology of culture : a critical reconstruction / Paul Jones.
p. cm.
Includes bibliographical references and index.
ISBN 0–333–66662–3 (cloth) — 0–230–00670–1 (pbk.)
1. Williams, Raymond. Culture. 2. Culture. I. Title.

HM621.J66 2003
306—dc21 2003056409

10 9 8 7 6 5 4 3 2 1
15 14 13 12 11 10 09 08 07 06

Printed and bound in Great Britain by
Antony Rowe Ltd, Chippenham and Eastbourne

For My Parents
Norma Jones
and
Robert Jones (1924–1975)

Contents

List of Tables and Figures

Tables

Figures

Preface to the Paperback Edition

This book focuses on the development and usefulness of the later project within Raymond Williams's work. He commonly called this a sociology of culture. It is underpinned by a theoretical position he variously called cultural materialism and/or social formalism. I date the emergence of this later project from about 1968 and its period of consolidation as the late 1970s. It reaches its most programmatic formulation in the 1980 book published as *Culture* in the UK and *The Sociology of Culture* in the USA. It was still in active development at the time of his death in 1988.

I have provided a more detailed account of this book's rationale and argument in the original preface, which follows. As I recount there, some situation of Williams's earlier work also became necessary.

One implication of this approach is worth underscoring here. As the secondary scholarship on Williams has expanded, it has become obvious that the significance of Williams's work is more than can be adequately contained in a single overview monograph. The duty of the very valuable overview monographs to cover so much ground often led to their treatments of elements of the later project being too compressed to convey fully its theoretical and methodological sophistication.

We seem to have entered a period of more specialized scholarship on Williams and also one of an increasing internationalization of that scholarship. This would appear to be the third monograph on Williams in a row to have been published by an author based outside Britain. Inevitably, differences of emphasis have emerged amongst those of us who read – and are reading – Williams within the context of nations of origin and cultural identities other than those within which he lived. For me this meant that it was all the more important to elaborate the later project with a member of an international reading public as the implied reader.

This edition has not been substantially revised from the hardcopy version published in 2004. Only major typographical and some other similar corrections have been made. While this book makes much use of previously unaddressed texts by Williams, the Bibliography is strictly one of works cited. My thanks to Jill Lake of Palgrave Macmillan in seeing this project through to paperback form. Finally, I wish to acknowledge the invaluable personal support of Catherine Waldby.

Paul Jones
Sydney
December, 2005

Preface: Looking Both Ways

> People have often asked me why, trained in literature and expressly in drama, making an ordinary career in writing and teaching dramatic history and analysis, I turned – *turned* – to what they would call sociology if they were quite sure I wouldn't be offended (some were sure the other way and I'm obliquely grateful to them). I could have said, debating the point, that Ruskin didn't turn from architecture to society... But I would prefer to speak for myself. I learned something from analysing drama which seemed to me effective not only as a way of seeing certain aspects of society but as a way of getting through to some of the fundamental conventions we group as society itself.... It was by looking both ways, at a stage and a text, and at a society enacted in them, that I thought I saw the significance of the enclosed room – the room on the stage – with its new metaphor of the fourth wall lifted – as at once a dramatic and social fact. (Williams, 1975, pp. 18–19 and 21)

This book started out from what seemed a straightforward set of propositions formed around the time of Raymond Williams's death in 1988: that Williams's intellectual legacy was more than his biography; that however worthy his exemplary life, that legacy too would die unless some programmatic retrieval was attempted; that Williams himself had not left anything as obvious as a monograph that objectivated his 'programme'.

I was already impatient with certain trends within the dominant secondary scholarship on Williams published prior to his death, especially within literary and cultural studies. There was a recurrent 'deep form' that seemed more appropriate to an entry in an encyclopaedia of popular music: that the early Williams scored the 'big hits' that had since been sampled by his betters and that the later Williams was either unreadable or confused. Even the claimed period of Williams's early heyday (1958–61) connoted the career of an Elvis Presley-like figure who had strayed too far from his 'roots'.

Accordingly, my initial plan was to retrieve what I started calling the 'mature' or 'late' project that that orthodoxy – or variants of it – had occluded. This study would not follow the perfectly reasonable but soon well-trodden path of revisiting Williams's monographs in serial succession. The mature project was to be the organizing principle.

However, I underestimated the weight of that 'early big hits' critical orthodoxy and was amazed to discover that it rested largely on a confusion of Williams with – or unconscious displacement by – the figure of Richard Hoggart, founder of the Birmingham Centre for Contemporary Cultural Studies (CCCS). Following Williams's own small joke in a book review – that hinted at a recognition of a kind of patronizing classism in that conflation – I called this 'the myth of "Raymond Hoggart"'.[1]

So contesting 'Raymond Hoggart' meant spending more time on those early works than I had intended but, for all that I draw from those works, the later project remains my chief concern. It is, in part, a rearticulation of some of Williams's earlier work but remains a distinctly coherent yet fragmented project that is arguably incomplete. It criss-crosses Williams's later writings from about 1969 and finds some of its most explicit expression in minor publications that have seen no republication and little subsequent discussion.

The most significant fragmentation within the late work in my view is the effective isolation of Williams's later writings on means of communication – including the better-known *Television* – from central texts like *Marxism and Literature* and *The Sociology of Culture*. Those central texts provide the overarching theoretical apparatus within which Williams located means of communication but they fail to discuss means of communication explicitly. As a consequence, even some of Williams's most enduring 'fellow travellers' – those often called 'political economists of the media' like Nicholas Garnham and Graham Murdock – have legitimately questioned Williams's eccentric prioritizations. It is for this reason that this book tends to build towards Chapter 6's reconstruction of what Murdock has called 'the absent centre' of Williams's mature project.[2]

But why call this reconstructed project a 'sociology of culture'? The easy answer is that the 'core' late text is the book published as *Culture* in Britain and *The Sociology of Culture* in the United States. I have found no better explanation for this name change than that implied by Bruce Robbins, that the book was published in an introductory sociology series in Britain and as a freestanding monograph in the United States.[3] Hence my use of the latter title throughout. In other words, it really was intended as a 'sociology of culture'. Once one continues from that premiss, a remarkable number of other jigsaw pieces lock into place. Williams appears to have been developing a social theory or sociology of culture since at least *The Long Revolution*.

Of course there is more to it than that. Let me add immediately, however, that my intention is not to claim Williams 'exclusively' for the discipline of sociology. As the citation from his 1974 inaugural professorial lecture that heads this preface makes plain, Williams was the least likely advocate of disciplinary boundaries in the faculty corridor sense and regularly advocated interdisciplinary work. But the late *research programme* he established is referred to as a sociological one with relentless frequency.

Let me try to clarify this. It's fairly well known that Williams stated three years after that inaugural lecture, that he had rejected literary criticism 'not only as an academic subject but as an intellectual discipline' (*WICTS*, p. 13). Plainly, however, he did not stop 'looking both ways', as he put it in the inaugural lecture. By 'looking both ways' he could recognize and avoid all forms of technicism, including both technological determinist approaches to means of communication and any 'anti-sociological' (his term) formalist analyses of cultural forms. His respect for the specificity of formalist analyses increased but was tempered by an insistence that they, too, should look both ways. When McLuhan, for example, argued that the formal specificity of 'the medium' was profoundly significant, Williams was intrigued; when McLuhan set aside all social determinants but 'the media' and 'projected' the affirmative utopia of the mediated global village, Williams attacked.

Looking both ways so entailed rethinking – and eventually typologizing – the existent and possible forms of relationship within and between 'society' and 'culture'. Williams's late project so provides a very explicit social critique of – and alternative to – the usual understandings of the linguistic and cultural 'turns'. He gives the resulting alternative its own name, *social formalism*, which I have adopted as the title of my reconstruction of this dimension in my fourth chapter. This is one of the least understood of his late self-positionings. Indeed, the opening passage reminds us that the linguistic and cultural turns in the social sciences were not all that far behind the 'sociological turn' in much of literary studies. However, Williams makes it explicit that orthodox sociology – understood as a naïvely empiricist functionalism and sometimes as a vulgar economism – has much to learn from social formalism as well. In particular, social formal analysis would need to be recognized as a component of empirical research; cultural modes like drama are 'social facts'.

Cultural materialism? That position certainly was announced – manifesto-like – by Williams in 1976 but it is an insufficient characterization of the late project in my view. I largely confine its use to the discussion in Chapters 2 and 3 of Williams's redefinition of his relationship with Marx and (Western) Marxism.[4]

This book does not directly address related contemporary debates in literary studies – for example, those around 'new historicism' – but aims instead to link Williams's work with comparable methodological and other literatures in sociology. Chapter 5 reconstructs Williams's (re)mapping of the sociological field as he understood it. It will undoubtedly seem strange to many sociologists. Yet it should at least dispel variants of the inverted Leavisism in sociological dismissals of Williams such as one from as recently as 1993 that argues his work 'is placed more accurately in a literary critical tradition'.[5]

And yet Williams's work on literary practices and forms *does* receive considerable discussion in this book. I seek to demonstrate in those and other discussions that Williams did indeed abandon the practice of literary criticism – understood as the close reading of Leavis's practical criticism and related discriminating judgements. However, rather than move towards the existent sociological orthodoxy, he embraced a practice of *emancipatory critique* that resembles most the critical sociological work of the Frankfurt School. Indeed the roots of this practice in Williams's work go back at least as far as *Culture and Society*. As I briefly discuss in the final chapter, it so happens that this practice also lies at the core of Habermas's public sphere thesis, with which others have legitimately associated Williams. Such *methodological* shifts – as Williams indicates in the passage above – provide a surer key to his later work than some remarkably misplaced accounts of 'his' conception of culture.

There is, however, a price Williams pays for his declared acts of theoretical clarification. Williams's later work tends to split his theoretical clarifications from his well-known emancipatory commitment to participatory democracy (and more). Emancipatory critique is practised by Williams but never receives quite the level of systematic recognition as other elements of his work. Accordingly my final chapter attempts to reconnect – to borrow a phrase from Adorno – these 'torn halves'.

I found it necessary to step outside some of Williams's self-characterizations elsewhere in the critical dimensions of this reconstruction as well. For instance, as I argue in Chapter 2, it is far more accurate to characterize the cultural materialism as the implementation of a 'production paradigm' in the field of 'culture'. Here, as György Márkus was the first to recognize, Williams's path closely resembles that of Adorno. It is this production paradigm that provides a bridge between the cultural materialism, the social formalism and the 'full' sociology of culture.

So perhaps the most direct challenge to much critical orthodoxy about Williams that this book offers is the degree to which I do link him with the project of Frankfurt Critical Theory. It is not well known that during his 'rapprochement' with the Marxian tradition Williams weighed up Marcuse, Adorno and Benjamin with a seriousness equal to that of his better-known assessments of Goldmann and Gramsci. It is from Goldmann, however, that Williams develops his sociology of formations: his analysis of self-organized aesthetico-intellectual groupings. This practice – first 'tested' in a critique of the Birmingham CCCS in 1977 – increasingly informs his work right up to his last writings on avant-gardism and cultural theory.

All of which might seem to place Williams at some alarming distance from cultural studies – despite his fading 'founding father' status therein. Rather, I see this book answering recent calls like Douglas Kellner's to redress 'the missed articulation' of (Frankfurt) Critical Theory and cultural studies.[6] As I argue in Chapter 1, it was an orthodoxy within cultural studies, rather, that undertook to distance itself from Williams many years ago. Yet, once one rejects 'the myth of Raymond Hoggart', very little of that orthodoxy is left standing, especially that concerning Williams's 'definitions of culture' that I examine in considerable detail. My own disappointment with the fate of the cultural studies project is undoubtedly evident at times in the following pages. The peculiar mishandling of Williams within its orthodoxy – most often undertaken, it must be added, in good faith – is undoubtedly one reason why this graduate of the CCCS 'turned' towards sociology. However, I hope my critical reconstruction of the limits of Williams's sociology of culture – most obviously his remarkable blindspots around issues related to popular culture and popular reception – might 'articulate' with some strands within contemporary cultural studies. That is one possible contemporary prospect for 'looking both ways'.[7]

Others have already criticized elements of the orthodoxies that this book aims to challenge.[8] Where I am aware of them, I have acknowledged these and other anticipations of – or debts owed by – my own arguments. Space considerations have required me to restrict to two 'excursuses' discussion of resemblances between the reconstructed sociology of culture and more recent research.[9]

Perhaps my hardest task – and the cause of greatest delays in completion – was deciding on what Marx called somewhere 'the mode of exegesis'. The more argumentative and citational evidence I provided for the positions I advanced about Williams, the less this book resembled the relative openness of Williams's better-known writings. For I also wanted the book to include the kind of 'user-friendly' introduction to his sociology of culture that, in my view, Williams had never provided himself. The compromise I reached was to employ an old sociological exegetical convention – the table. While they by no means condense my whole argument, the twenty odd tables lay out some 'core' elements of Williams's project. I also use them as a form of internal cross-referencing of the book's case. The tables might be especially useful to those who employ the book as a resource for their own critical research. At least I hope that is the case as the encouragement of such research was one of my main motivations in writing this book.

Paul Jones
Sydney
March, 2003

Acknowledgements

Any extended process of scholarly research and writing incurs many personal debts. This project has been far more extended than anticipated so the debts are considerable.

My immediate teaching colleagues at UNSW School of Sociology have been unflaggingly supportive over many years. During that time – as a group and perhaps 'formation' – we lived through an horrific confirmation of Williams's prescient warning that critical sociology is one of the most fragile of autonomous intellectual spaces: Jocelyn Pixley, Clive Kessler (especially for those *Brumaire* discussions), Maria Márkus, Michael Pusey, Michael Bittman, Mira Crouch and, all too briefly of late, David Holmes.

György Márkus's influence on this project will be obvious to the reader. Many hours of conversation over many years have helped shape key arguments.

Other friends and colleagues who have provided intellectual support include Michael Symonds, Hart Cohen, Pauline Johnson, John Grumley, Andrew Milner, John Rundell, Jennifer Wilkinson, Judy Wajcman and Craig Browne.

I am also very grateful for discussions with those who, unlike me, knew Williams as a personal colleague: Nicholas Garnham, John Fekete, Francis Mulhern, Graham Martin and, some years ago now but still memorably, Stuart Hall. Sections of this book forcefully question Hall's interpretation of Williams but his early formative influence on the shaping of this project is also considerable. Nicholas Garnham encouraged me far more than he realized when he commented that I was 'worrying away at the right questions'. For different reasons, I also thank my 'musical' conversations with my fellow CCCS alumnus, Andrew Goodwin, and with Dave Laing; and for her very encouraging correspondence, Janet Wolff. My thanks too to Graeme Turner for his gracious feedback on my brief critique in Chapter 1 of his characterization of Williams's place within cultural studies.

Arguments developed for the book have benefited from discussions at many conference and seminar sessions. Notable amongst these for me were those at the Department of Sociological Studies, University of Sheffield, and the Centre for Communication and Information Studies, University of Westminster, in 1996, and the remarkable *Raymond Williams: After 2000* conference organized by Andrew Milner and *Overland* in Melbourne in July 2000.

I gained valuable interdisciplinary UNSW Faculty feedback while teaching related materials with Martyn Lyons, Damian Grace, Olaf Reinhardt, Ludmila Stern, Alan Krell and Stephen Gregory. This book has benefited from UNSW Arts Faculty grants which enabled me to fund Olaf Reinhardt's and Kerry Burgess's translation assistance and the research assistance of Ian Andrews, Kerry Burgess, Wai Chan, Nigel Smith and Liz Turnbull. I also thank Nigel

Smith for posing his characteristically no-nonsense Brummie question, 'Just what is Williams's method, anyway?', during a course I taught in 1991. Likewise thanks to Saadi Nikro for drawing me back towards *Modern Tragedy* at the right moment. My thanks to similar contributions that came from many undergraduates and postgraduates in the course of teaching and supervision.

Special thanks to Denise Thompson for reading the entire manuscript – lately heroically under considerable time constraints – and providing razor-sharp constructive commentary. My thanks also to Nikó Antalffy.

My publisher, Palgrave Macmillan, has been patient well beyond the usual necessary tolerance of the pace of scholarly 'productivity'. Special thanks to Tim Farmiloe right at the beginning and, more recently, Jennifer Nelson and Catherine Gray.

Sections of this book have previously appeared in earlier forms in *The Canadian Journal of Communication, Cultural Studies, KeyWords* and *Culture and Enlightenment: essays for György Márkus*. Publication details are listed under my name in the bibliography.

Acknowledgement is made of permissions to use extracts from the following works by Raymond Williams: *Culture and Society* published by Chatto and Windus used by permission of the Random House Group Limited; *Culture and Society* (1958) used by permission of the US publisher, Columbia University Press; *The Long Revolution* published by Chatto and Windus used by permission of the Random House Group Limited; *What I Came to Say* published by Hutchinson used by permission of the Random House Group Limited; *The Sociology of Culture* published by Fontana used by permission of the Random House Group Limited; *The Sociology of Culture*, copyright © 1981 used by permission of the US publisher, Shocken Books, a division of Random House Inc.; *Marxism and Literature* reprinted by permission of Oxford University Press; *Politics and Letters* used by permission of Verso Press.

The author and publisher have made every effort to identify copyright holders. If any have been inadvertently overlooked the appropriate arrangements will be made at the first opportunity.

Abbreviations of Titles and Editions of Williams's Books Cited

All other cited works of Williams are entered alphabetically in the Bibliography. For a complete listing of Williams's publications see Alan O'Connor's bibliography in O'Connor (1989) and Eagleton (1989b).

C&C	*The Country and the City*	(1975) London: Paladin. First published by Chatto & Windus, 1973
C&S	*Culture and Society: Coleridge to Orwell*	(1990) London: The Hogarth Press. '1987 edition'. First published as *Culture and Society: 1780–1950* by Chatto and Windus, 1958
COM1	*Communications* (1st Edn)	(1962) Harmondsworth: Penguin
COM2	*Communications* (2nd Edn)	(1966) Harmondsworth: Penguin
COM3	*Communications* (3rd Edn)	(1976) Harmondsworth: Penguin
DFIE	*Drama From Ibsen to Eliot*	(1965) London: Chatto & Windus. First published 1952
DIP	*Drama in Performance* (2nd Edn)	(1972) Harmondsworth: Penguin. First published 1968. First Edn 1954
KW1	*Keywords* (1st Edn)	(1976) London: Fontana
KW2	*Keywords* (2nd Edn)	(1983) London: Flamingo
LR	*The Long Revolution*	(1965) Harmondsworth: Pelican. First published by Chatto & Windus, 1961
M&L	*Marxism and Literature*	(1977) Oxford: Oxford UP
MT1	*Modern Tragedy* (1st Edn)	(1966) London: Chatto & Windus
MT2	*Modern Tragedy* (2nd Edn)	(1979) London: Verso (Restructured edition with new afterword)
O	*Orwell* (3rd Edn)	(1991) London: Fontana First edition published 1971
P&L	*Politics and Letters: Interviews with New Left Review*	(1979) London: New Left Books
PMC	*Problems in Materialism and Culture*	(1980) London: Verso
POM	*The Politics of Modernism: Against the New Conformists*	(1989) London: Verso

1979	*Politics and Letters: interviews with New Left Review*
	'Afterword to *Modern Tragedy*'
1980	*Problems in Materialism and Culture*
1981	*The Sociology of Culture*
	Contact: human communication and its history (ed.)
1983	*Towards 2000*
	'Marx on Culture'
1984	*Writing in Society*
1986	'The Uses of Cultural Theory'
1987	'Language and the Avant-Garde'
1988	'The Politics of the Avant-Garde'
1989	*The Politics of Modernism: against the new conformists*

1
Settling Accounts with 'Culture'

1.1 Preliminaries: culture is ordinary?

Williams's 'expansive' usage of the category of culture is the achievement for which he is most widely known. Certainly, references to his 'definition of culture' are the most common form of citation of his work. The expansion usually attributed to Williams seems quite straightforward: from a narrowly aesthetic confinement to a widened 'anthropological' reach, culture as 'a whole way of life'. One of Williams's most famous phrases from his early work – taken from a 1958 essay of the same name – seems to sum this perspective up: 'Culture is Ordinary'.[1]

The central problem with this emphasis as a way of approaching Williams's work is that it disembeds his usage of 'culture' from the context of his own arguments, and especially his emancipatory normative criteria. The success of his historical semantic vocabulary book, *Keywords*, has facilitated this emphasis. While *Keywords*'s entries are hardly 'objective', they do not necessarily reveal Williams's own position. *Keywords* cannot be made to stand for Williams's project(s).

'Culture is ordinary' is especially misleading if divorced from Williams's own usage. It seems to invite a reversal of its subject and object, so implying an *indiscriminate* 'equalization' of all artefacts: the ordinary is culture. But Williams plainly intended the paradoxicality of the original formulation. As Francis Mulhern has recently suggested, it is better understood as 'creation is ordinary' (Mulhern, 2000, p. 81). This passage from the 1958 essay points towards the fuller complexity of Williams's position:

> We use the word culture in these two senses: to mean a whole way of life – the common meanings; to mean the arts and learning – the special processes of discovery and creative effort. Some writers reserve the word for one or other of these senses; I insist on both, and on the significance of their conjunction. (*ROH*, p. 4)

'Creation is ordinary' might then be seen as the consequence of this deliberate 'conjunction' by Williams. On this reading, all humans are capable of creative practice and so culture is thus rendered ordinary. The way in which Williams draws on his working-class background to demonstrate this point in the same essay is thus highly significant:

> At home we met and made music, listened to it, recited and listened to poems, valued fine language. I have heard better music and better poems since; there is the world to draw on. But I know, from the most ordinary experience, that the interest is there, the capacity is there. (*ROH*, pp. 5–6)

The 'ordinary experience' of these practitioners included demonstrated cultural creativity within an awareness and reception of 'fine' and 'better' arts. And learning? Williams provides the concrete example of his father's autodidactic acquirement of critical information which, as Williams junior puts it, 'had had made easy for me in two or three academic essays' (*ROH*, p. 13).

Williams's 'culture is ordinary' thus resembles Gramsci's more famous phrase, 'all men are intellectuals' (Gramsci, 1971, p. 8). Mulhern points us towards a passage in *The Long Revolution* with uncanny resemblances to that discussion in Gramsci:

> The suggestion that art and culture are ordinary provokes quite hysterical denials, although, with every claim that they are essentially extraordinary, the exclusion and hostility that are complained of are in practice reinforced. The solution is not to pull art down to the level of other social activity as this is habitually conceived. The emphasis that matters is that there are, essentially, no "ordinary" activities, if by "ordinary" we mean the absence of creative interpretation and effort. (*LR*, p. 54)

But even this 'Gramscian' corrective misses Williams's persistent use, in these same discussions, of terms like 'learning' and 'effort' and the rejection of the implication of a 'levelling' in the 'quality' of cultural creativity. Indeed, Williams's *Long Revolution* discussion continues to redefine 'the arts' as 'learned human skills, which must be known and practised in a community before their great power in conveying experience can be used and developed' (*LR*, p. 54). Williams made it plain that he was decidedly not abandoning qualitative judgement. 'Culture is Ordinary' and *Communications* challenge, respectively, 'the observable badness of so much widely distributed popular culture' and the 'many kinds of routine art and routine thinking', while rejecting the retreat into conceptions of 'minority/mass' or 'high/low' binarizations of culture as inadequate responses to this acknowledged problem (*ROH*, p. 12; *COM1*, p. 72).

Williams later provided his own clarification of the 'significance of the conjunction' of the two senses of culture above:[2]

> It was . . . as a way of exploring an alternative emphasis, of discovering a standpoint within this complex territory, that one tried to speak of a common culture, or (the phrase now seems to me different) a culture in common. Related to this stress was the assertion that culture is ordinary: that there is not a special class, or group of men, who are involved in the creation of meanings and values, either in a general sense or in specific art and belief. Such creation could not be reserved to a minority, however gifted, and was not, even in practice, so reserved . . . In talking of a common culture, then, one was saying first that culture was the way of life of a people, as well as the vital and indispensable contributions of specially gifted and identifiable persons, and one was using the idea of the *common* element of the culture – its community – as a way of criticizing that divided and fragmented culture we actually have.
>
> It was . . . perfectly clear that the majority of people, while living *as* people, creating their own values, were both shut out by the nature of the educational system from access to the full range of meanings of their predecessors in that place, and excluded by the whole structure of communications – the character of its material ownership, its limiting social assumptions – from any adequate participation in the process of developing meanings which was in any case going on. One was therefore both affirming a general truth, which I would hold to be independent of any particular historical stage, that there is, in that sense, community of culture; and criticizing a particular society because it limited, and in many ways actively prevented, that community's self-realization. (*ROH*, pp. 34–5)

Moreover, Williams goes on to explicitly identify that 'self-realization' with an '*educated and participating democracy*' (*ROH*, p. 37).

In sum, much of the confusion related to Williams's usage of 'culture' derives from his commitment to these two lesser known criteria in 'Culture is Ordinary':

(a) His dual-purpose use of the category and related formulations as a means of constructing a critical-emancipatory social norm *and* as a means of 'empirical' assessment against that norm.
(b) The second overlapping criterion derives from his view that while cultural 'democratization' did entail the rejection of the élitism of the various minority/mass formulations, it did *not* entail 'equalizing' all existing cultural acts as if they were of equal qualitative aesthetic merit.

Williams's early usage of terms like 'bad culture' could thus indicate a failure to meet either or both of these criteria. These criteria emerged as part of Williams's engagement with contemporary 'English' debates about the expansion of educational opportunity in general and, more particularly, the teaching of *skills* of critical interpretation and reception, especially as a response to that 'routinization' evident in both 'high and low' culture. As this educational commitment never leaves Williams, it is important to assess how his normative positions differed from his precursors like Leavis and Eliot, and especially from his contemporary, Richard Hoggart, best known as the author of *The Uses of Literacy* and founder of the Birmingham CCCS. These differences will be discussed in the next section, and their implications for cultural studies orthodoxy in the following section.

Of course the theoretical and normative grounding of the early Williams's qualitative judgements of 'bad culture' underwent radical reassessment and redefinition – but never abandonment – in his subsequent work. This chapter thus also opens discussion of Williams's complex journey of methodological and normative enquiry as it develops within his early and some of his later discussions of 'culture'.

1.2 Against class reductivism and a mythologized 'organic community'

> At the time when Richard Hoggart and I were inseparable, we had not yet met. It still seems reasonable that so many people put his *Uses of Literacy* and my *Culture and Society* together. One newspaper went (s)o far as to refer, seriously, to a book called *The Uses of Culture* by Raymond Hoggart. But as I say we did not then know each other, and as writers we were pretty clear about our differences as well as our obvious common ground. (Williams, 1970a)

Even as it stands, Williams's statement above immediately discredits the anecdotal belief that *The Uses of Literacy* and *Culture and Society* constituted in the period 1957–58 a co-ordinated assault on an élitist orthodoxy by two of the 'angry young men', that formation of intellectuals then prominent in the British public life.[3] They simply had not met and knew each other only through correspondence and publications.

Yet the two were indeed both scholarship boys from working-class backgrounds in Wales and Leeds who were trained as Leavisite literary critics. Williams's and Hoggart's closer commonality was their criticism of Leavis's intended social role for intellectuals such as themselves, 'a saving minority' of cultural missionaries.

Leavis's plan was a contemporary revision of a project initiated in the English case by Samuel Coleridge and developed by Matthew Arnold – the establishment of a stratum of state-provided cultural intellectuals, a *clerisy*.[4] It was the proposed social practices of a cultural clerisy that Williams and Hoggart separately challenged.

Leavis radically restricted the appropriate qualifications for a twentieth century clerisy member to that of a literary critic. Coleridge's clerisy, by contrast, had included 'all the so-called literal arts and sciences' (Coleridge, 1852, p. 55); while Arnold had socially grounded his 'culture' more firmly by his blasé assumptions about what 'the great men of culture', acting with the exemplary disinterestedness of 'sweetness and light', had constituted as 'the *best* knowledge and thought of the time' (Arnold, 1971, p. 70).[5]

Leavis's rationale for his further restriction of ('minority') 'culture' to (select) literature is quite fundamental. Only certain literary traditions provide a link with the lost 'organic community' where 'the picked experience of ages' was deposited in folk traditions and craft skills:

> And such traditions are for the most part dead.... It now becomes plain why it is of so great importance to keep the literary tradition alive. For if language tends to be debased... instead of invigorated by contemporary use, then it is to literature alone, where its subtlest and finest use is preserved, that we can look with any hope of keeping in touch with our spiritual tradition – with the "picked experience of ages". But the literary tradition is alive only so long as there is a tradition of taste, kept alive by the educated (who are not to be identified with any social class); such a tradition – the "picked experience of ages" – as constitutes a surer taste than any individual can pretend to. (Leavis and Thompson, 1937, p. 82)

This is a major source of Leavis's 'moral values' embedded in 'Literature' that are retrievable by the 'close reading' of his 'practical criticism'. Likewise, the process of 'keeping the literary tradition alive' is primarily one of cultivation of appropriate taste criteria amongst literary consumers. But this is not a simplistic defence of 'high culture' for its own sake. Rather it is a strategy socially premissed on a radically critical minority confronting 'mass civilization'. The ambition of broader cultural renewal remains, but now within that consumptive limitation to literary consumers. The social dispersal of the practice of critical-consumptive 'scrutiny' outside a narrow intelligentsia constitutes an implicit political programme. Leavis's oppositional formation could aim for quite radical cultural objectives, such as the abolition of advertising.[6] These are nonetheless co-present with the fundamental conservativism of the mass/minority dichotomy which Leavis adds to the clerisist critique of utilitarian 'civilization'.

Crucially, however, this ongoing conservativism means that the creative spontaneity of the folk (now 'masses'), a major Rousseauian assumption of much Romantic thinking, is abandoned. This is claimed to be 'debased', indeed dead. In Leavis, clerisism becomes explicitly linked to a denial of popular capacity for productive creative practice.[7]

It was the arrogance of this more limited élitism that Williams and Hoggart both rejected. Both had become de facto members of Leavis's clerisy as extramural teachers to adult, usually working class, students. Following Leavis's own example they applied the 'close reading' of his practical criticism to both canonical literature and popular cultural material. Both wrote textbooks developed from this practice.[8] Each found his own trajectory to be a denial, not a confirmation, of Leavis's premisses about the social distribution of creative capacity. Hoggart's textbook provides this neat summation of his dilemma:

> But one's misgivings are not so much about the method itself as about the spirit in which it is sometimes advocated. There is too often a calvinistic self-righteousness of manner and a bloodless intellectualism which may be proper to the training of an "intellectual saving minority" but is an unsuitable frame of mind in which to approach the special problems of adult students.... Our students' response to experience is often much richer and more courageous than we at first suspect. We should base our work on this fine capacity; we should aim more at encouraging and developing what is already there, instead of behaving like an anti-tetanus team in a primitive community. (Hoggart, 1963, p. 9)

This recognition of 'fine creative capacity' is certainly common ground with 'Culture is Ordinary'. The recognition of 'what is already there' provides an entry into the strategy of *The Uses of Literacy* – the book's representation of the cultural life of the contemporary British working class as something other than the degraded consumption of mass cultural commodities. As is now well known, Hoggart establishes a case for the existence of a 'way of life' culture based primarily within the social relations of working-class inner-city neighbourhoods. He also teases out remarkable nuances in his case studies of 'oblique attention' in *reception* – the non-passive or non-designed usages of popular cultural commodities. Such insights revealed a depth of familiarity, albeit nostalgic, which was simply beyond the social reach of Leavis's work and a considerable influence on the Birmingham agenda.[9]

Williams's commentary on Leavis in *Culture and Society* makes his reasons for the unacceptability of Leavis's 'organic community' thesis quite explicit:

> This is, I think, a surrender to a characteristically industrialist, or urban, nostalgia – a late version of mediaevalism, with its attachments to an "adjusted" feudal society. If there is one thing certain about "the organic community", it is that it has always gone. Its period, in the contemporary myth, is the rural eighteenth century; but for Goldsmith, in *The Deserted Village* (1770), it had gone;... for Cobbett, in 1820, it had gone since his boyhood... for myself (if I may be permitted to add this, for I was born into a village, and into a family of many generations of farm labourers) it was there – or the aspects quoted, the inherited skills of

> work, the slow traditional talk, the continuity of work and leisure – in the 1930s.... it is foolish and dangerous to exclude from the so-called organic society the penury, the petty tyranny, the disease and mortality, the ignorance and frustrated intelligence which were also among its ingredients. These are not material disadvantages to be set against spiritual advantages; the one thing that such a community teaches is that life is whole and continuous – it is the whole complex that matters. (*C&S*, pp. 259–60)

The Uses of Literacy was published in 1957, a year before *Culture and Society*. Williams published two reviews of it and subjected its central category of 'working-class culture' to an 'immanent critique' in the conclusion of *Culture and Society*.[10]

Williams finds in Hoggart too an over-dependence on the conservative dimensions of clerisism that he identified in Leavis and charted in detail in his own book. The following passage is aimed squarely at the mass/minority dichotomy but also demonstrates the necessary disjunction Williams sees between Hoggart's lingering dependence on clerisism and the responsibility of intellectuals from working-class backgrounds:

> The analysis of Sunday newspapers and crime stories and romances is of course familiar, but, when you have come yourself from their apparent public, when you recognize in yourself the ties that still bind you, you cannot be satisfied with the older formula: enlightened minority, degraded mass. You know how bad most "popular culture" is, but you know also that the irruption of the "swinish multitude", which Burke prophesied would trample down light and learning, is the coming to relative power and relative justice of your own people, whom you could not if you tried desert. My own estimate of this difficulty is that it is first in the field of ideas, the received formulas, that scrutiny is necessary and the approach to settlement possible. Hoggart, I think, has taken over too many of the formulas, in his concentration on a different kind of evidence. He writes at times in the terms of Matthew Arnold, though he is not Arnold nor was meant to be. (*WICTS*, p. 26)

But Williams reserves his harshest criticism for one of the components of Hoggart's category of working-class culture:

> Finally, he has admitted (though with apologies and partial disclaimers) the extremely damaging and quite untrue identification of "popular culture" (commercial newspapers, magazines, entertainments etc.) with "working class culture". In fact the main source of this "popular culture" lies outside the working class altogether, for it was instituted, financed and operated by the *bourgeoisie*, and remains typically capitalist in its

> methods of production and distribution. That working class people form perhaps a majority of the consumers of this material, along with considerable sections of other classes..., does not, as a fact, justify this facile identification. In all of these matters, Hoggart's argument needs radical revision. (*WICTS*, p. 27)

'Culture is Ordinary' launches a swingeing attack on the 'cheapjacks' of what today would be called 'tabloid culture'. In a passage in *Communications* that he kept in its three editions published between 1962 and 1976, Williams condemns a synthetic 'anti-culture' as 'not the culture of "the ordinary man"; it is the culture of the disinherited' (*COM1*, p. 74; *COM2*, p. 102; *COM3*, p. 115). Plainly, Williams saw in education the prospect of a 'reinheritance' of the 'common inheritance' and of skills lost to a population that nonetheless still held creative capacity.

Williams continues the critique of Hoggart (without naming him) within the conclusion to *Culture and Society*. There, after reiterating the disjunction between popular culture and control of its production, he continues:

> "working class culture", in our society, is not to be understood as the small amount of "proletarian" writing and art which exists. The appearance of such work has been useful, not only in its more self-conscious forms, but also in such material as the post-Industrial ballads, which were worth collecting. We need to be aware of this work, but it is to be seen as a valuable dissident element *rather than as a culture*. The traditional popular culture of England was, if not annihilated, at least fragmented and weakened by the dislocations of the Industrial Revolution. What is left, with what in the new conditions has been newly made, is small in quantity and narrow in range. It exacts respect, but is in no sense an alternative culture. (*C&S*, p. 320; emphasis added)

Here we can see the source of Williams's continuing problems with the category of 'popular culture' – especially the growing significance of popular music – and the methodological option of ethnography. A non-commercial popular culture is only recognized briefly in *Communications* and more forcefully in *The Country and the City* and *Towards 2000*.[11] While Williams rejected Leavis's organic community thesis *tout court*, he appears to have derived his position about the fate of the 'traditional popular culture of England' directly from him.[12]

As Georgina Boyes has recently argued, the key influence here is that of Cecil Sharp's version of Romantic folkloricism (and its role in the contemporary English Folk Revival) upon Leavis's 'lost organic culture' thesis.[13] Most especially, one crucial collection of notated folksongs published by Sharp, *English Folk Songs from the Southern Appalachians*, had recounted the remarkable maintenance of the English (and Scottish) folksong tradition amongst the Appalachian communities of the USA. Sharp goes to great lengths in his introduction to stress the uniqueness of the Appalachians' 'way of life', most especially their apparent prioritization of leisure time and especially singing over material comforts.[14] For Leavis, the survival of this folksong tradition demonstrated the necessary integration of authentic folk culture within an organic community, a ' "way of life" (in our democratic parlance) that was truly an art of social living' (Leavis, 1966, p. 190). Its negation was the 'mass civilization' of contemporary England. As Leavis saw salvation in the teaching of literature, Sharp saw it in his highly disciplined programme of teaching his approved curriculum of folksong and, especially, folkdance. Williams criticized Sharp explicitly in 1973 in *The Country and City*. Although the following revised understanding of 'the post-Industrial ballads' plainly meets the later Williams's criteria for an 'authentic popular culture', he would almost certainly have characterized the ballads as a 'residual' rather than an 'alternative' cultural form:

> There was the abstract and limiting definition of folksong which in Cecil Sharp was based on the full rural myth of the "remnants" of the "peasantry", and which specifically excluded, as not of the "folk", the persistent songs of the industrial and urban working people, who did not fit the image but were continuing to create, *in an authentic popular culture*, what it suited this period and this class to pretend was a lost world. (*C&C*, p. 309; emphasis added)

In *Culture and Society's* critique of Leavis, Williams similarly advocates the relevance of 'other experience' 'more various than literature alone'. He cites not only other forms of 'recorded culture' – effectively restoring Coleridge's 'all the sciences' – but also 'experience that is otherwise recorded: in institutions, manners, customs, family memories' (*C&S*, p. 255). In the next section of this chapter Williams's 'historicist' method in *The Long Revolution* is reconstructed but even the similar inclusiveness of its 'documentary' conception of culture does not follow up the ethnographic implication of that second list in the critique of Leavis. Nor does the issue of ethnography arise later in Williams's work.[15]

Williams's prioritization of an 'alternative culture', however, was maintained and was to be reformulated within his later embrace of the concept of hegemony.[16] In *Culture and Society* it is tentatively positioned against 'bourgeois culture'. But the latter too requires reformulation. Williams insists that 'the body of intellectual and imaginative work which each generation receives as its traditional culture' is always 'a common inheritance'. Class interest thus manifests in the transmission and distribution of 'the common inheritance', especially by means of the mechanism of the 'selective tradition'. Thus: 'The manufacture of an artificial "working class culture", in opposition to this common tradition, is merely foolish' (*C&S*, pp. 320–1).

This comment refers to the earlier rejected definition of working-class culture as a marginalized 'proletarian writing'. But Williams also continues to develop the complexity of his account of the class–culture relationship:

> If we think of culture, as it is important to do, in terms of a body of intellectual and imaginative work, we can see that with the extension of education the distribution of this culture is becoming more even, and, at the same time, new work is being addressed to a public wider than a single class. Yet a culture is not only a body of intellectual and imaginative work; it is also and essentially a whole way of life. The basis of a distinction between bourgeois and working class culture is only secondarily in the field of intellectual and imaginative work, and even here it is complicated, as we have seen, by the common elements resting on a common language. The primary distinction is to be sought in the whole way of life, and here, again, we must not confine ourselves to such evidence as housing, dress and modes of leisure. Industrial production tends to produce uniformity in such matters, but the vital distinction lies at a different level. The crucial distinguishing element in English life since the Industrial Revolution is not in language, not dress, not leisure – for these will tend to uniformity. The crucial distinction is between alternative ideas of the nature of social relationship. (*C&S*, p. 325)

Here we can see Williams struggling to reconcile the two senses of 'culture' we met at the beginning of this chapter. The alternatives are presented as two distinct operational conceptions of the relation between human individuals and society. The bourgeois alternative of individualism which leaves 'society' as a neutral field of action for competition is pitted against the working-class alternative of solidarity which sees society as 'the positive means for all kinds of development, including individual development'. The latter leads to the following famous (re)definition:

> We may now see what is properly meant by "working class culture". It is not proletarian art, or council houses, or a particular use of language; it is, rather, the basic collective idea, and the institutions, manners, habits of thought and intentions which proceed from this. Bourgeois culture, similarly, is the basic individualist idea, and the institutions, manners, habits of thought and intentions which proceed from that. In our culture as a whole, there is both a constant interaction between these ways of life and an area which can properly be assigned as common to or underlying both. The working class, because of its position, has not, since the Industrial Revolution, produced a culture in the narrower sense. The culture which it has produced, and which it is important to recognize, is the collective democratic institution, whether in the trade unions, the cooperative movement or a political party. Working class culture, in the stage through

> which it has been passing, is primarily social (in that it has created institutions) rather than individual (in particular intellectual or imaginative work.) When it is considered in this context, it can be seen as a very remarkable creative achievement. (*C&S*, p. 327)

As we have seen, 'Culture is Ordinary' too builds on Williams's own central proposition of culture as a 'common inheritance'. It manages to establish all the central tenets of the conclusion to *Culture and Society* without once using the category of 'working-class culture'. That subcategory is never actively or systematically employed in Williams's work again.[17]

Nonetheless, it is important to stress how open was Williams's alternative – from which he developed his many communications and cultural policy proposals.[18] It is based in his conception of common inheritance and the open education of cultural 'skills':

> Nobody can raise anybody else's cultural standard. The most that can be done is to transmit the skills, which are not personal but general human property, and at the same time to give people open access to all that has been made and done. (*C&S*, pp. 318–19)

But 'whole way of life' is a different matter. It cannot be set aside as easily as 'working-class culture' because it endures within Williams's own practice. Accordingly, it offers an even more convincing rhetorical figure than 'culture is ordinary' to support the view that Williams moved *from* aesthetic 'high culture' *to* a relativizing anthropological understanding as a 'whole way of life'. It is true that in his very earliest discussions of the concept of culture, Williams sourced the 'whole way of life' meaning to sociology and anthropology.[19] With considerable prescience he remarks that this usage is 'likely to cause confusion' but finds it is necessitated because of a key step taken by the members of his English 'culture and society' tradition: 'the extension of a critic's activities in the judgement of works of art to the study and thence the judgement of "a whole way of life" ' (1953, p. 240).

The phrase, 'whole way of life', was drawn from T.S. Eliot's *Notes Towards the Definition of Culture*. It was Eliot, not Williams, who first employed the phrase as a rendering of an 'anthropological' sense of culture within 'critical judgement' extended beyond works of art.[20]

Eliot introduces the 'whole way of life' sense of culture in somewhat extraordinary circumstances. The varying clerisy proposals had offered culture as an emulation of religion and so as a court of appeal against industrial capitalism's perceived excesses. Eliot simply reverses this secular assumption by asserting that culture is the incarnation of the religion of a people. It is thus religion, not culture, that Eliot initially proposes as 'the *whole way of life* of a people' (Eliot, 1948, p. 31).[21] Yet he also wishes to include within religion a behavioural 'lived' dimension – from 'culture' – that is broader than the Christian emphasis on religion as belief. It is in order to demonstrate this broader reach of culture that he introduces his famous miscellany of English cultural activities: 'Derby Day, Henley Regatta, Cowes, the twelfth of August, a cup final, the dog races, the pin table, the dart board, Wensleydale cheese, boiled cabbage cut into sections, beetroot in vinegar, nineteenth-century Gothic churches and the music of Elgar' (Eliot, 1948, p. 31).

Williams correctly notes that this miscellany is merely a playful rendering of Eliot's more serious point. Yet Williams is also correct to point out that it is in such moments of play that Eliot allows his definitions to slide, for example between 'arts and learning' and 'whole way of life' (*C&S*, p. 234). Accordingly, Williams's critique aims to lay bare the serious conclusions Eliot draws from his sliding 'definitions'. For Eliot also argues, in explicit opposition to all forms of egalitarianism in education, that only an élite dedicated to the maintenance of a 'conscious' culture can maintain successful cultural transmission. Here, crucially, he reminds his reader that culture is 'not merely the sum of several activities' (an apparent allusion to his miscellany) 'but a *way of life*'. This is Eliot's most surreptitious sleight of hand as a cultural élite is not only necessary for Eliot but must be grounded within a *way of life*. The only appropriate way of life available, as it happens, is that of the dominant social class.[22]

It is hardly surprising that Williams initially stated, with uncharacteristic bluntness for his work in this period, that Eliot's book 'is a work almost calculated to infuriate' (Williams, 1956, p. 307).[23] Yet Eliot's provocative formulations also provided evidence for Williams's case against class reductivism and so helped him formulate his emphasis on a common culture and its means of transmission. For Eliot did pit his conception of culture as 'the creation of the society as a whole' against Karl Mannheim's definition of culture as the product of an intelligentsia within his theory of merit-based élites.[24]

Eliot's engagement with Mannheim enables Williams to place Mannheim within the culture and society tradition and also to begin to draw his case in *Culture and Society* to a close:

> Mannheim's argument may be seen, fundamentally, as an epilogue to the long nineteenth century attempt to reidentify class with function. This took the form, either of an attempt to revive obsolete classes (as in Coleridge's idea of a clerisy), or of an appeal to existing classes to resume their functions (Carlyle, Ruskin), or of an attempt to form a new class, the civilizing minority (Arnold). Mannheim, quite rightly, realizes that these attempts have largely failed. Further, he rejects the idea of classes based on birth or money, and, emphasizing the necessary specialization and complexity of modern society, proposes to substitute for the old classes the new élites, whose basis is neither birth nor money, but achievement. (*C&S*, p. 239)

Williams rejects Eliot's conservative insistence on a maintained governing social class, but also mobilizes Eliot's hostility to Mannheim's élite(s) as a 'refinement of social *laissez-faire*' (*C&S*, p. 240). Eliot's conservative critique thus exposes the failings, not only of Mannheim, but also of 'the ordinary social-democratic case' and 'orthodox "liberalism"' (*C&S*, p. 241).

If one rejects Eliot's own solutions, Williams argues, then 'the next step must be in a different direction, for Eliot has closed almost all the existing roads' (*C&S*, p. 243). Thus is the scene set for Williams's closing chapter on 'Marxism and Culture' and for his conclusion.

Throughout this critique Williams increasingly adopts and transforms Eliot's 'whole way of life'; for example: 'the definition of culture as "a whole way of life" is vital at this point for Eliot is quite right to point out that to limit, or attempt to limit, the transmission of culture to a system of formal education is to limit a whole way of life to certain specialisms' (*C&S*, p. 240).

It is possible that the Mannheim/Eliot contrast enabled Williams to set up his 'arts and learning versus whole way of life' contrast in 'Culture is Ordinary'. Even his attempted solution to this problem of reconciliation in conceptions of cultural skill emerges here.[25]

To reiterate, of these phrases – culture is ordinary, working-class culture, culture as whole way of life – only the last survives into Williams's later work.[26] However, it does so *not* as a reference to an 'anthropological flattening', but as a reference to Eliot's posing of the problem of how to theorize more adequately the connections between the social relations of cultural creation/production and 'transmission' on the one hand, and 'the arts and learning' on the other. For Williams these questions are initially resolved, as we saw, in the normatively charged model of a 'common culture' to which we shall return later in this chapter.

1.3 Problems of 'Culturalism' or 'Cambridge'?: cultural studies parts company with Williams

The complexity of Williams's adoption of 'whole way of life' has been reconstructed in some detail because of a highly influential misinterpretation that lies at the heart of the cultural studies project. Likewise, the differences between Williams and Hoggart have gone largely unrecognized within the project which grew from the centre Hoggart established at Birmingham.

Such confusions may also explain why the cultural studies claim to Williams as a 'founding father' has been at most ambivalent. Hoggart's successor as Director at Birmingham, Stuart Hall, provided a famous paradigmatic characterization of the field in 1980 as a competition between 'culturalist' and 'structuralist' paradigms.[27] This has since consolidated into an orthodoxy best exemplified by Graeme Turner's textbook introduction to the field, in which the culturalism/structuralism binary is sequentially narrativized with Williams located with Hoggart within the former culturalist phase.[28] Turner acknowledges Williams's 'definitions' as foundational for the field of cultural studies, but laments the lack of a corresponding methodological legacy in his work. This 'methodological absence' might have been filled, for Turner, by structuralist semiotics. The possibility that this mismatch is the product of a misconstrual of Williams's 'definitions' does not arise. In one very revealing formulation, Turner states that Williams 'founds a tradition that others develop' (Turner, 1996, p. 54).

In one sense this last comment is true. Williams's mature work is not part of what is now recognized, at least in orthodox accounts like Turner's, as 'cultural studies'. We see the legacy of "Raymond Hoggart" at work within cultural studies more explicitly in the work of Stuart Hall and his successor at Birmingham, Richard Johnson. As in Turner's later summary account, in this process Williams's work is conflated with Hoggart's as *definitionally* foundational, then set aside for various kinds of alleged theoretical inadequacies.

Even by 1980 it may have been possible to argue that Williams's cultural materialism was still underdeveloped. However, as recently as 1997 Hall reiterated his critique's basic premisses in an interview.[29] In 1993 he introduced his most recent elaboration of this critique – which always takes the form of a 'break thesis' modelled on Althusser's account of Marx's intellectual development[30] – thus:

> In his discussion of culture, in the famous chapter on "The analysis of culture" in *The Long Revolution*, his pathbreaking attempt to break with the literary-moral discourse of *Culture and Society* into a more sustained effort of general theorizing, the key conceptual move he makes is from an "abstract" definition of culture – "a state or process of human perfection" – to culture as "a description of a particular way of life which expresses certain meanings and values not only in art and learning but in institutions and ordinary behaviour". Culture, he insisted, with his characteristic inflection on "our common life" is "ordinary". The analysis of culture, from such a definition, he argued, "is the clarification of the meanings and values implicit and explicit in a particular way of life, a particular culture". Characteristic here is not only the movement from abstract ideal to concrete, from texts to their contexts of institutional life and ordinary behaviour; but also the breaking down of artificial distinctions between art and literature – the signifiers of "culture" in the first, as it were "Cambridge" sense – and what he called "the general social organization". (Hall, 1993, p. 351)

This break thesis is unsustainable. According to Hall, this break occurred sometime between the publication of *Culture and Society* in 1958 and *The Long Revolution* in 1961. On his own account, the phrase 'culture is ordinary' sums up the 'later' position. Yet the article of that name was published in

1958, the same year as the 'pre-break' *Culture and Society*. That article anticipates many of the more overtly empirical sociological dimensions of *The Long Revolution*. Any close examination of the two books, especially in relation to contemporaneous articles, suggests strongly that they were published by 'the same author' but merely had different foci. Indeed, Williams retrospectively described their composition as virtually a joint process.[31] In the introduction to *The Long Revolution*, he notes the continuity between elements of the two books and regards the completion of both as the ending of 'a stage of my life' (*LR*, p. 15).

It follows from this, either that *Culture and Society* does not operate in some anterior 'literary-moral discourse', or that Williams saw a continuing role for some such critical practice. Hall's use of the phrase, 'anterior discourse', resembles the proposition put forward by Perry Anderson in 1968 that Leavisite literary criticism filled the role of an 'absent sociology' within the development of twentieth-century British intellectual culture (Anderson, 1968). This 'anterior discourse' so enabled the emergence of *The Long Revolution*.[32] An Althusserian assumption is detectable – arguably more so in Hall than Anderson – that a break from a 'pre-scientific' moral discourse to 'general theorizing' is a necessary step in theoretical clarification. For Althusser such discernment of the correct 'theoretical object' is a key part of this process.[33] This, I suggest, is the source of Hall's and others' fascination with Williams's definitions and redefinitions of 'culture'.

Anderson's thesis about *The Long Revolution* in turn bears some resemblance to that advanced seven years earlier by E.P. Thompson about *Culture and Society*: 'With a compromised tradition at his back, and a broken vocabulary in his hands, he did the only thing that was left to him: he took over the vocabulary of his opponents, followed them into the heart of their own arguments and fought them to a standstill in their own terms' (Thompson, 1961, p. 27).

The ambiguities and clumsiness of some of these war-weary formulations are, as Williams later conceded, quite evident. However, as Hall and others continue to (re)circulate them without adequate reference to their conditions of composition or Williams's later work, precision in the reconstruction of these early texts and contexts becomes all the more essential.

Yet even Thompson's more sophisticated understanding of Williams's early strategy needs supplementation. Williams did deliberately research the *Culture and Society* 'tradition' as a counter-tradition, especially against Eliot, and certainly followed his opponents 'to the heart of their own arguments' but he went further than the stalemate implied by Thompson's 'standstill'.[34] Thompson's critique set a template for later critics of Williams from the left: that Williams's analysis was somehow politically compromised by his detailed engagement with those he criticized in *Culture and Society*. However, as we saw in the critique of Hoggart, the final stage of Williams's critique was to provide an alternative 'content' for elements of his opponent's 'vocabulary'.

But even if we set aside this issue (until the next section) and confine ourselves to Hall's emphasis on definitions, there is a still more serious problem for his characterization. The section of *The Long Revolution* on which Hall relies discusses *three*, not two, meanings of 'culture'.[35] Given the continuing confusion about this text, a detailed reconstruction of its argument is warranted here. Williams lays out three approaches, a corresponding mode of analysis for each and a range of possible methods following from these modes (Table 1.1).[36]

Hall claims Williams rejects definition (i) and embraces (iii) without even acknowledging Williams's presentation of (ii). As can be easily seen from the table, this grossly misrepresents Williams's typologization. Hall's total elision of (ii) is very significant for, as I will argue, it informs Williams's preferred practice. In the original 'Two Paradigms' discussion, Hall acknowledges that 'way of life' 'has been rather too neatly abstracted' from Williams's text but he still reduces Williams's typologization to two definitions by conflating (ii) with (iii) and equating the resulting documentary/anthropological with 'ethnographic' methods (Hall, 1980a, p. 59). Yet Williams never mentions such methods!

For just a moment Hall acknowledges Williams's key step of 'reconciling' more than one definition by arguing that he (Williams) 'integrated' the way of life into the 'central' ideal one. However, while this comes closer to acknowledging Williams's stated position, it is also part of Hall's means of distancing himself from Williams's alleged 'culturalism'.

Hall sourced the charge of culturalism to Richard Johnson but its coinage occurred, ironically, in one of the first defences of Williams, by Anthony Barnett, against Terry Eagleton's 1976 critique.[37] Johnson, however, provided perhaps the most pivotal (re)formulation of the charge of culturalism:

> As literary critic and cultural theorist, Williams does stress certain kinds of practices, all of them broadly cultural, and, within that, mainly literary. Other practices tend to be marginalized or defined away. There is no check on this from theoretical controls. Thus the early works are particularly inattentive to political processes, a tendency which Williams himself has acknowledged.[38] The tensionless "expansion" of culture replaces struggle over values and definitions. Though some of this is repaired in later work, there is a persistent neglect of the particular character and force of economic relations and therefore of economic definitions in relation to class. This "culturalism" is described by Anthony Barnett, the most careful of Williams's critics, as a kind of inversion of economism, a reduction "upwards". This is the characteristic tendency of 1950s and 1960s texts in both history and "literary sociology". It is very characteristic of Hoggart's *The Uses of Literacy*, for example, from which both economic production and politics are literally absent. (Johnson, 1979a, p. 218)

Table 1.1 *The Long Revolution's* preliminary typologization of 'The Analysis of Culture'[a]

Definition of 'culture'		Analysis of culture which follows from this definition	Possible methodological range within such an analysis
(i) Ideal	'A state or process of human perfection'	'The discovery and description in lives and works, of those values which can be seen to compose a timeless order'	None provided
(ii) Documentary	'The body of intellectual and imaginative work in which, in a detailed way, human thought and experience are variously recorded'	'The activity of criticism, by which the nature of the thought and experience, the details of the language form and convention in which these are active, are described and valued'	From (Arnoldian) *ideal criticism* that focusses on a particular work – 'its clarification and valuation being the principal end in view' to *historical criticism* 'which, after analysis of particular works, seeks to relate them to the particular traditions and societies in which they appeared'
(iii) Social	'A description of a particular way of life, which expresses certain meanings and values not only in art and learning but also institutions and ordinary behaviour'	'The clarification of the meanings and values implicit and explicit in a particular way of life, a particular culture'	From *historical criticism* (as above) to the (sociological) analysis of arguably 'extra-cultural' elements: organization of production, structure of the family, structure of institutions, characteristic forms of communication[b]

[a] All citations from *LR*, pp. 57–8.
[b] Williams adds that such analysis displays the same methodological range as the cultural, but it is unclear whether he means here all the interests of 'social analysis' he lists or only the last relating to communication.

Three theses would appear to be present in this argument:

(a) that Williams does not 'set boundaries' to the concept of culture and fails to define it as more than 'way of life';

(b) that as 'literary critic and cultural theorist' Williams marginalized other practices, especially political and economic practices;
(c) that this culturalism is typical of the 'literary sociologies' of the 1950s and 1960s, the best example of which is Hoggart's *The Uses of Literacy*.

Within this argument are valid observations. For example, Williams does indeed privilege literary practices in his analyses of aesthetic culture; and it is true that he conceded that he displaced concepts such as the state in his work, as he assumed they were adequately theorized by others.[39] But, as we have seen, he hardly marginalized politics or economic conceptions of class. The crucial step, however, is the last in which Johnson substitutes Hoggart for Williams. In fact, as we saw, Williams criticized Hoggart on issues (a) and (b) himself.

It is also significant that Johnson declares earlier that the main problem with 'the culture problematic' is that, as the 'tradition was an overwhelmingly literary one, the debate was evaluative rather than analytic' (Johnson, 1979a, p. 212). The apparent rejection of the former for the latter is perhaps the most characteristic feature of the charge of 'culturalism' and what most marks it as Althusserian. As we have seen, it is also the element that Hall has continued to promote.

It is true that Williams's work contains an undeclared mode of critique of socio-cultural works and forms which is normative (rather than 'evaluative'), and which emerges as that feature of his project most vulnerable to continuing attempts to maintain the myth of 'Raymond Hoggart'. This acknowledgement, however, begs the somewhat repressed question in Williams scholarship: what then is the relationship between his 'evaluative' literary analyses and his 'other' analyses?

Hall and Johnson tended to see Williams's evaluative analyses as a 'literary-moral discourse' from which he needed to 'break'. It is certainly valid then for Hall to regard Williams as having broken with a 'Cambridge' cultural legacy, and Turner is correct in identifying a certain ambiguity in Williams's methodological legacy. Williams did retrospectively admit (in 1977) to breaking from the evaluative discourse of 'Cambridge literary criticism' 'as an intellectual discipline' (*WICTS*, p. 13).

But what did he break *towards*? The answer to this 'question of method' is immensely complex and provides a major motif for this book. But we can find the beginnings of an answer by examining in more detail the contents of Table 1.1.

1.4 'This is a problem of method...'[40]

Hall's claims about the alleged 'break' Williams makes in *The Long Revolution* result from inadequately contextualized citations from this passage:

> I would then define the theory of culture as the study of relationships between elements in a whole way of life. The analysis of culture is the attempt to discover the nature of the organization which is the complex of these relationships. Analysis of particular works or institutions is, in this context, analysis of their essential kind of organization, the relationships which works or institutions embody as parts of the organization as a whole. A key-word, in such analysis, is pattern: it is with the discovery of patterns of a characteristic kind that any useful cultural analysis begins, and it is with the relationships between these patterns, which sometimes reveal unexpected identities and correspondences in hitherto separately considered activities, sometimes again reveal discontinuities of an unexpected kind, that general cultural analysis is concerned. (*LR*, p. 63)

The influence of the critique of Eliot is fairly obvious here. Between this passage and the options outlined in Table 1.1, Williams employs a case study in order to 'test' all the listed methods. It is the same example used by Eliot in his *Notes*, the 'clash of duties' in Sophocles' *Antigone*.[41] Let us again resort to a Table (1.2) as an aid.

Williams spends considerable time pointing to the failings of the tendency of the social-contextual method in particular to reduce the results of (i) and (ii) to 'contexts to which we have assigned them' (*LR*, p. 60). (Remember that this is the approach Hall claims Williams advocates!) While such socio-cultural contextualization gains much over an abstract-ideal method, it risks producing, like the ideal method, a categorical bifurcation of art and society. Such a bifurcation can be overcome, Williams insists,

Table 1.2 Cultural analyses of Sophocles' *Antigone*

Definition of 'culture'	**Implications for analysis of Sophocles' *Antigone***	**Result**
(i) Ideal	Derivation of a timeless 'ideal value'	Recognition of ideal value of reverence for the dead
(ii) Documentary	Communication of certain values by certain means	Basic human tensions conveyed by dramatic form of chorus and double *kommos*, and the specific intensity of the verse
(iii) Social	Recognition of limitations of the context of the particular culture within which the play was produced and so provided the 'timeless values' derived in (i); Recognition of social shaping of specific dramatic forms in (ii)	Antigone's preparedness to die for her brother's right to burial, a product of the specific kinship system of ancient Greece

only by theoretical acknowledgement of the mutual embeddedness of 'all the activities and their interrelations, without any concession of priority to any one of them we may choose to abstract' (*LR*, p. 62). Here Williams reintroduces the *documentary* mode of analysis 'because it can yield specific evidence about the whole organization' (*LR*, p. 62). He then begins to build a model of an adequate 'cultural history' in which 'particular histories' are returned to 'the whole organization'. At this point of his argument the long statement above is made, clearly referring to the bringing together of these disparate 'elements'.

As with the critique of Hoggart, the proto-conceptual language here is undoubtedly tortuous but once the context is restored, Williams's strategy becomes clearer. For in returning to and privileging the documentary sense of culture, he is evidently exploring the other method listed against it in Table 1.1, an *historical criticism*.

Indeed his immediately following statement makes this clear:

> It is only in our own time and place that we can expect to know, in any substantial way, the general organization. We can learn a great deal of the life of other places and times, but certain elements, it seems to me, will always be irrecoverable. Even those that can be recovered are recovered in abstraction, and this is of crucial importance. We learn each element as a precipitate, but in the living experience of the time every element was in solution, an inseparable part of the complex whole. The most difficult thing to get hold of, in studying any past period, is this felt sense of the quality of life at a particular place and time: a sense of the ways in which the particular activities combined into a way of thinking and living. (*LR*, p. 63)

Williams rejects Fromm's 'social character' and Benedict's 'pattern of culture' as inadequate to this task and so introduces his notoriously difficult 'structure of feeling', initially defined within that same tortuous language as 'the particular living result of all the elements of the organization'.[42] Just prior to this he draws an analogy that is more revealing – that this phenomenon can be detected within the reader's contemporary culture in the shifts in meaning of particular words recognized by different generations.[43] This not only points to the method of *Keywords* but the rationale for writing the book with which it was intended to be published, *Culture and Society*.[44] Finally, Williams also links 'structure of feeling' with the role of the arts '(f)or here, if anywhere, this characteristic is likely to be expressed' (*LR*, p. 65). Likewise, a generational shift in structures of feeling is the moment that may provide greatest access to 'the whole organization'.

By this point in his argument, 'the documentary culture' is clearly Williams's key 'object' of analysis as it provides the best access to the structure(s) of feeling. Williams stresses that the documents, 'from poems to buildings and dress fashions', are not necessarily autonomous:

> It is simply that, as previously argued, the significance of an activity must be sought in terms of the whole organization, which is more than the sum of its separable parts. What we are looking for, always, is the actual life that the whole organization is there to express. The significance of documentary culture is that, more clearly than anything else, it expresses that life to us in direct terms, when the living witnesses are silent. (*LR*, p. 65)

At this point Williams introduces another more famous formulation:

> We need to distinguish three levels of culture, even in its most general definition. There is the lived culture of a particular time and place, only fully accessible to those living in that time and place. There is the recorded culture, of every kind, from art to the most everyday facts: the culture of a period. There is also, as the factor connecting lived culture and period cultures, the culture of the selective tradition.
>
> One can say with confidence, for example, that nobody really knows the nineteenth-century novel; nobody has read, or could have read, all its examples, over the whole range from printed volumes to penny serials.... Equally, of course, no nineteenth-century reader would have read all the novels; no individual in the society would have known more than a selection of its facts. But everyone living in the period would have had something which, I have argued, no later individual can wholly recover: that sense of the life within which the novels were written, and which we now approach through our selection. Theoretically, a period is recorded; in practice, this record is absorbed into a selective tradition; and both are different from the culture as lived. (*LR*, pp. 66–7)

Obviously this argument contests the notion that an existent literary canon, for example, should be the only point of entry into what we saw was Leavis's 'picked experience of ages'. Yet while Williams brings to theoretical recognition an awareness of the mechanism of the selective tradition, he acknowledges the inevitability of some such process. We have also seen that he explicitly rejects Leavis's celebration of a mythical organic community that some such texts are deemed to embody.[45] Crucially, then, there is no evidence here that Williams valorizes this historically irrecoverable 'lived experience' as more authentic than the present (as might Leavis). His historicism is more radical than that. Rather, the primary role of the 'level' of lived experience here is to underscore the selectivity of the selective tradition. Also, if we briefly summarize this model figuratively (Figure 1.1), something else becomes obvious.

This 'recorded culture' from which the tradition is selected is obviously identical to 'documentary culture'. This provides a methodological answer to the question of how Williams 'democratized' his conception of culture:

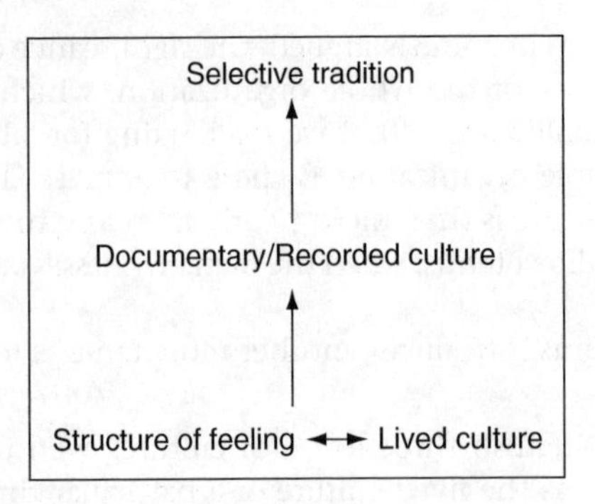

Figure 1.1 *The Long Revolution*'s 'three levels of culture'.

not by rejecting 'high culture' for 'low culture' or for an anthropological conception of 'way of life culture' but by admitting *all objectivated culture* as 'documentary culture' to the first stage of the reconstruction of a structure of feeling.

Williams afterwards provides his much admired demonstration of this historicist historical critical method – his case study of the structure of feeling of the 1840s.[46] It can be presented as four stages:

(i) reversal of the literary selective tradition in order to generate a fuller field of documentary culture
(ii) location of this documentary culture within economic and technical changes within cultural institutions
(iii) location of these in turn within 'the general social and political history of the period'
(iv) establishment of links across these three fields using the concepts of 'social character' and 'structure of feeling'.

This sequence strongly informs the need for an examination of socio-cultural institutions in *The Long Revolution* and the mature sociology of culture and so includes a reformulated social definition of culture. Stage (iv) warrants further expansion. Fromm's 'social character' is employed as an effective replacement for the class-based 'alternative ideas of the nature of social relationship' introduced in the conclusion to *Culture and Society*. The 1840s thus provide a kind of historical testing ground as a case study of the contest of class ideals presented there. The bourgeois social character is established as the dominant one via 'its characteristic legislation, the terms in which it was argued, the majority content of public writing, and the characters of the men most admired' (*LR*, p. 78).[47] It is subject to contestation by similarly derived aristocratic and working-class social characters. The ideal of public service – considered and personally rejected for that of solidarity, in the conclusion to *Culture and Society*[48] – is seen to be a product of such contestation.

The structure of feeling, while often coincident with the social character, also 'has to deal not only with the public ideals but with their omissions and consequences' (*LR*, p. 80). It is in exploration of these contradictory dimensions that Williams examines the documentary culture. He regards the popular periodical fiction as enacting the bourgeois social character quite directly. However, its assertion 'that success followed effort' presents narrative challenges when confronted by the plausible eventuality of failure. The narrative reliance on 'magical devices' such as unexpected fortune reveals these points of tension between the social character and the structure of feeling. The structure of feeling is so rendered analytically visible. Moreover, 'the structure of feeling as described... is present in almost all the novels we now read as literature, as well as in the now disregarded popular fiction' (*LR*, p. 84). However, whereas the popular fiction routinely employs the magical devices to resolve contradictions, the literature registers them through a range of methods (including the magical devices) that are often more self-conscious. These, Williams says, 'are the creative elements' that allow even conventional forms to register 'a radical human dissent' (*LR*, p. 85). Williams so concludes:

> Art reflects its society and works a social character through to its reality in experience. But also art creates, by new perceptions and responses, elements which the society, as such, is not able to realize. If we compare art with its society, we find a series of real relationships showing its deep and central connexions with the rest of the general life. We find description, discussion, exposition through plot, and experience of the social character. We find also, in certain characteristic forms and devices, evidences of the deadlocks and unsolved problems of the society: often admitted to consciousness for the first time in this way. Part of this evidence will show false consciousness, designed to prevent any substantial recognition; part again a deep desire, as yet uncharted, to move beyond this. (*LR*, p. 86)

The method endorsed in 'The Analysis of Culture', then, relies on the missing 'third option' overlooked by Hall's and others' accounts, an historical criticism based initially in the *documentary* conception of culture which later reincludes a reformulated social definition (cf. Table 1.1). This mode of analysis examines a 'documentary culture' in conjunction with a critique of the 'organization' of the selective tradition later developed from it. Yet within the analysis that follows from this 'documentary' inclusiveness, as we have seen, the arts are effectively privileged. There is no 'anthropological' flattening into an undifferentiated 'way of life' conception of culture and no inclusion of ethnographic methods. Nor, crucially, is there an *a priori* abstraction of an aesthetic 'high' culture. This exploratory role for art assumes an integral relation of art within society. Williams's reason for this ongoing privileging of the arts follows from the normative dimension articulated above, rather than any defensive opposition to 'mass culture'. But these norms are not the same as the ideals of the 'ideal conception of culture' Williams typologizes. It is the critically reflective capacities of autonomous cultural practice that points to *unrealized possibilities* – wherever they might be located – that Williams alternatively valorizes.

This almost entirely anticipates the method Williams was to deploy in *Modern Tragedy* (and to which he refers therein as 'historical criticism'). In perhaps the strongest parallel with his historical semantic analysis of 'culture', he acknowledges the evident tension between the orthodox scholarly and popular understandings of 'tragedy'. This mismatch does not require a mere subordination of the latter to the former nor, of course, an endorsement of the scholarly complaint that the popular applications of 'tragedy' – to, for example, car accidents – constitute misuses of the word. Rather, the current scholarly understanding is revealed to be, like 'culture', the product of a selective tradition. Williams's reconstruction of this tradition emphasizes the new social content that was invested in the (European) dramatic mode as it survived beyond its origins within the collectively embedded metaphysical assumptions of Greek tragedy. But modern tragedy *per se* – most characteristically in Ibsen's 'liberal tragedy' – is chiefly concerned, as in the *The Long Revolution* analysis above, with 'the deadlocks and unsolved problems of society', that is, the failure to pursue 'unrealized possibilities'.

Even within *The Long Revolution's* case studies, Williams does not confine himself to structures of feeling in the past. The book moves on through its seven socio-historical case studies of cultural forms and cultural institutions and finally builds to a significant culmination: the analysis of another decade, not another lost 'lived experience' but the *coming* decade, 'Britain in the sixties'.[49] In 1983 Williams republished 'Britain in the sixties' in its entirety in the equally future-focussed *Towards 2000* and then subjected it to 'reconsideration' and 'extension' as a form of prospective analysis.[50]

In the introduction to *The Long Revolution* Williams retrospectively identifies his critical analysis of 'structures of feeling' with the chapter on 'The Industrial Novels' from *Culture and Society*.[51] That analysis indeed resembles the analysis elaborated above.[52] But if 'The Analysis of Culture' leads to a theoretically reflective reconstruction of the method deployed in *only one* of the chapters of *Culture and Society*, what method was Williams deploying in the rest of it?

The key to this hidden method lies within the mode by which Williams redefines the ideal conception of culture.

1.5 Williams's undeclared method: immanent critique

> Everywhere in the nineteenth century we see men running for cover from the consequences of their own beliefs. (*MT1*, p. 70) (1965)

Williams notes in *Politics and Letters* that even as a student he made subversive revisions to the 'close reading' of Leavis's 'practical criticism':

> The normal *Scrutiny* practice in the criticism of fiction was to judge the quality of a novel or of a novelist by analysing a sample of prose which was assumed to be a representative pattern of the writer's work as a whole. This method was developed essentially for the analysis of the single short poem. I didn't think it would work with the novel. Already in preparing for the Tripos I searched for a long time to find paired examples of prose by George Eliot and Lawrence that would demonstrate the point. The cases I chose showed that one pair would make George Eliot a better writer than Lawrence, and the other pair would make Lawrence a better writer than George Eliot. At the time I felt this to be a challenge to the critical orthodoxy. (*P&L*, p. 237)

The point of this subversive tactic was to demonstrate the arbitrariness of the *Scrutiny* 'proof' of judgement and the need for a more 'wholistic' engagement with the text (a position Williams later abandoned). Of course in these cases Williams was addressing fictional writing. But as Dan O'Neill has recently detailed, the extension of the close reading of practical criticism to non-fictional prose is one of the unacknowledged pathbreaking steps taken by Williams.[53] Yet the above passage also draws attention to the number of 'paired examples' *of writers* Williams employs in *Culture and Society*'s 'nineteenth-century tradition': Burke/Cobbett; Southey/Owen; Mill on Bentham and Coleridge; Newman/Arnold; even Eliot/Mannheim. More fundamentally, however, Williams's close readings of these authors seek *internal contradictions* in their *argumentative* prose. That is the 'subversive' reworking of practical criticism that Williams brings to argumentative prose. For reasons which will become clearer in Chapter 3, the term 'immanent critique' shall continue to be used for this practice.

It was argued in Section 1.3 that Williams's mode of critique in *Culture and Society* does not stop at bringing his opponents' arguments 'to a standstill' but, rather, continues on to reconstruct their 'content'. The 'content' of his reconstruction of the ideal conception of culture emerges most clearly in his critique of Matthew Arnold in *Culture and Society*.

As we have seen, Arnold too contributed to the legacy of Coleridge's clerisy model. He proposed that the appropriate social agent of cultural reconstruction was an ostensibly 'disinterested', and effectively transcendent, state.

His analysis in *Culture and Anarchy* (1869) is famous for its 'disinterested' definition of culture at the conclusion of the chapter on 'Sweetness and Light':

> culture works differently. It does not try to teach down to the level of the inferior classes; it does not try to win them for this or that sect of its own, with ready-made judgements and watchwords. It seeks to do away with classes; to make the best that has been thought and known in the world current everywhere; to make all men live in an atmosphere of sweetness and light, where they may use ideas, as it uses them itself, freely, – nourished and not bound by them.
>
> This is the *social idea*; and the men of culture are the true apostles of equality. The great men of culture are those who have had a passion for diffusing, for making prevail, for carrying from one end of society to the other, the best knowledge, the best ideas of their time; who have laboured to divest knowledge of all that was harsh, uncouth, difficult, abstract, professional, exclusive; to humanize it, to make it efficient outside the clique of the cultivated and learned, yet still remaining the *best* knowledge and thought of the time, and a true source, therefore, of sweetness and light. (Arnold, 1971, p. 70)

Arnold searches among existing social classes for suitable bearers of what he calls 'cultural authority', that is, those possessing exemplary characteristics rendering them suitable for the holding of state power as the basis for the cultural dissemination of disinterestedness. Arnold finds no single class suitable. He is particularly harsh in his estimation of the working-class's potential for such a role. Williams intervenes sharply to correct his assessment.

'Anarchy' is the most potent contemporary obstacle, in Arnold's analysis, to his goals. The maintenance of a national stability during the campaigns for democracy is Arnold's own socially anchored role for culture. Sweetness and light to this extent are not 'disinterested' ideals. The organized working-class's campaign for suffrage is consistently seen in the book to be the potential anarchic threat, as Williams cites in this passage:

> for us, – who believe in right reason, in the duty and possibility of extricating and elevating our best self, in the progress of humanity towards perfection, – for us, the framework of society, that theatre on which this august drama has to unroll itself, is sacred; *and whoever administers it*, and however we may seek to remove them from the tenure of their administration, yet, while they administer, we steadily and with undivided heart support them in repressing anarchy and disorder; because without order there can be no society, and without society there can be no human perfection. (Arnold, 1971, pp. 202–3)[54]

After this citation, Williams immediately commences his immanent critique:

> It is here, at so vital a point, that we see Arnold surrendering to a "stock notion or habit" of his class. The organizing, and at times demonstrating, working class was not, on any showing, seeking to destroy society as such. It was seeking by such methods as were available to it, to change the particular ordering of society which then prevailed....
>
> For Arnold to confuse the particular, temporary ordering of interests, which was indeed being threatened, with human society as such, is the confusion which elsewhere he so clearly analysed.... When the emphasis on State power is so great, any confusion between that ideal State which is the agent of perfection, and this actual State which embodies particular powers and interests, becomes dangerous and really disabling. (*C&S*, pp. 124–5)

The first phase of the critique is thus completed. Arnold's position in the above is seen to be vulnerable to an immanent application of his own preferred principles. This echoes the thesis Williams develops in his above-mentioned analysis in the previous chapter of *Culture and Society*, 'The Industrial Novels' – that a *fear of violence* pervaded the upper and middle classes in the period of democratic reforms and that it acted 'as an arresting and controlling factor' in intellectual work (*C&S*, pp. 90ff). When Williams continues his critique of Arnold, he builds from this thesis to a major introduction of a key element of his own programme:

> The case is one which Arnold, detached from his particular position, would readily understand. A prejudice overcomes "right reason", and a deep emotional fear darkens the light. It is there in his words: *hoot, bawl, threaten, rough, smash*. This is not the language of "a stream of fresh thought", nor is the process it represents any kind of "delicacy and flexibility of thinking". Calm, Arnold rightly argued, was necessary. But now the Hyde Park railings were down, and it was not Arnold's best self which rose at the sight of them. Certainly he feared a general breakdown, into violence and anarchy, but the most remarkable facts about the British working-class movement, since its origins in the Industrial Revolution, are its conscious and deliberate abstention from general violence and its firm faith in other methods of advance. These characteristics of the British working class have not always been welcome to its more romantic advocates, but they are a real human strength, and a precious inheritance. For it has been, always, a positive attitude: the product not of cowardice and not of apathy, but of moral conviction. I think it had more to offer to the "pursuit of perfection" than Matthew Arnold, seeing only his magnified image of the Rough, was able to realize. (*C&S*, p. 125)

This remains Williams's chief mode of rhetorical address for the remainder of *Culture and Society* and, indeed, for most of his writing until the mid-1960s. Arnold's 'breakdown in his thinking' provides Williams with the opportunity to constitute the 'moral conviction' of the working class as the more appropriate motor of the 'pursuit of perfection' and so the means of realization of 'the tradition' of English commentators on 'culture'.

However, the almost irresistible temptation in the post-Burkean Romantic critique was the surreptitious social anchoring of the perfectibility process, most obviously in particular artworks. Williams concludes of Arnold in *Culture and Society* that his slippage into 'the best that has been thought and written in the world' results in the relativization of his absolute criterion of human perfection and culture. This exposed the impossibility of its functioning as *both* the abstracted absolute constituted as critical court of appeal *and* the process of the desired reordering of existent socio-cultural institutions. For all the Romantics' efforts, the clerisy ideal lacked an acknowledged (as opposed to surreptitious) social context for its own advocated process. That it was a programme of democratic reform which socially derailed the views of one of its key advocates was thus no coincidence for Williams.

In a remarkably prescient formulation, Williams notes:

> Culture was a process, but he (Arnold) could not find the material of that process, either, with any confidence, in the society of his own day, or, fully, in a recognition of an order that transcended human society. (*C&S*, p. 127)

Here is perhaps Williams's first sketch of the need for a 'cultural materialism'. *The Long Revolution* provided, to Williams's satisfaction, an initial means of identifying the key 'material of the process' that required institutional and, indeed, societal, reorganization. Williams's initial articulation of his 'long revolution' is thus the proposed process of completion of this push for democratization towards and by means of which the 'long revolution' is moving.

Clearly there is a linkage between this critique of Arnold and the contemporaneous critique of Hoggart. What Williams proffers here as an historical 'alternative content' for Arnold's culture he also proffers for Hoggart's contemporary notion of working-class culture: the democratizing influence of the institutions of the organized working class.

The only recognition of Williams's practice of immanent critique as such appears to have come in Perry Anderson's 'Origins of the Present Crisis' in 1964.[55] Anderson pinpointed the technique's limits as a mode of politico-conjunctural analysis – that the 'positivity' Williams attributed in his institutional definition of working-class culture lacks 'a distinction between corporate and hegemonic institutional forms' (Anderson, 1964, p. 44). That is, as Williams would later put it, these institutions could become incorporated into the existing social order and so contingently lose their role as exemplary alternatives. Anderson's early (for English language writers) invocation of Gramsci so set the agenda very precisely for Williams's 1973 'Base and Superstructure' essay.[56]

On Williams's own account, the 'end of the road' for this pre-Gramscian 'positivity' came with the failings of the first and especially second Wilson Labour governments (1964–66; 1966–70)[57] – in industrial relations, their approach towards the Vietnam war and in cultural policy.[58] Williams co-authored *The May Day Manifesto* with Thompson and Hall in 1967–68, and by 1969 was drawing explicit parallels between contemporary opposition to anti-Vietnam demonstrations in London's Grosvenor Square and Arnold's criticisms of the 1866 Hyde Park demonstrations for the suffrage.[59]

Yet while Williams could no longer so readily provide an alternative 'content' in his immanent critiques, his *immanent mode of critique* was maintained and developed. Indeed, his continuing practical commitment to such *critique* stands in stark contrast to his growing hostility to (literary) 'criticism'.

Here then is a significant anomaly in the early Williams. He goes to tortuous lengths to draw his readers to his method of historical criticism for the analysis of structures of feeling and provides a whole 'theory of culture' largely to this end. Yet, outside his recognition of his reconstruction of a normative conception of culture, he does not reflect at all on his practice of immanent critique. Both these techniques can be seen as radical transformations of the practical criticism in which he was trained. Together they provide an adequate characterization of his initial solution to his 'problem of method'. Both seek to unlock the 'unrealized possibilities' present within autonomous culture. The underelaboration of immanent critique is partly redressed in Williams's later reflections on 'culture', discussed in the next section, and his engagement with the 'Western Marxists' discussed in Chapter 3.

1.6 Post-Romantic Enlightenment: later formulations of 'culture'

Both *Culture and Society* and *The Long Revolution* clearly assumed a British readership. The former's 'tradition' of authors is a British one while the latter's social histories of cultural institutions and policy proposals are prospectively addressed to a British polity – 'Britain in the sixties' – still deemed capable of radical social change. Likewise, the historical scope of each was deliberately confined: *Culture and Society* by its subtitular '1780–1950', and *The Long Revolution* by its titular emulation, in part, of the industrial revolution. And yet Williams was keen to remind his British readers that his three revolutions – democratic, industrial and cultural – were incomplete largely because their benefits were so confined to 'the advanced countries' (*LR*, pp. 10–11).

In any case, this 'localism' changes dramatically in Williams's later comparable writings on 'culture'. As he recomposes his assessment of the concept, his sources broaden to include (largely European) 'non-British' perspectives, extending to include especially the legacy of the Enlightenment.

Williams's discussion of 'culture' in *Marxism and Literature* and the little known essay, 'On High and Popular Culture', provides a good guide to his mature assessment of the concept. Characteristically, he sees the semantic history of 'culture' and its subsidiaries as the key.[60]

The chief gain for Williams is a more explicit understanding of the Enlightenment ideal of culture and the Romantic critique thereof, most notably that undertaken by Herder, the late eighteenth-century German writer and folklorist. It is Herder who first cogently articulates the Romantic critique of the dominant Enlightenment conception of culture (*kultur*).

The Enlightenment usage principally signified the progressive process of secular human self-development or 'self-making'. It was to find its fullest articulation in the development of philosophical aesthetics in works such as those of Schiller and Hegel. It was thus closely related to the central Enlightenment category of *reason*. One useful summation of this conception is the following: 'culture is the process of developing and ennobling the human faculties, a process facilitated by the assimilation of works of scholarship and art and linked to the progressive character of the modern era' (Thompson, 1990, p. 126).

Implicit in this 'facilitation' is the overlap with the related, and at times interchangeable, Enlightenment category of *civilization*. A tension was present in each between the ideal/abstract and empirical/concrete forms of this process. While both categories embodied for the Enlightenment the abstracted human potentiality for *ongoing* self-development throughout history, the pull towards anchoring their meanings in a contemporary set of materialized achievements was constant. For civilization this anchoring could be in courtly 'manners' (and later the national State); for culture, such anchoring was primarily in objectivated artworks and the cultivation of an individual's 'taste'.

The most familiar anchoring in popular English usage today is the related word 'cultivated', which bears more strongly the limited pedagogical goal of individual development. But 'cultured' and 'civilized' also still signify the limitation that was central for Williams: the celebration of an already achieved state rather than the immanent potential of a progressive social process. Williams traces the complex semantic mutations of these categories in terms of this 'problematic double sense' of ideal and anchored meanings (*M&L*, p. 14).

The specific difference in the British case, however, was that 'culture' also came to signify a *reversal* of the progressive developmental dimensions of the Enlightenment project rather than a Herderian Romantic qualification. Arnold's blasé assumptions about 'the best', Eliot's anti-secular redefinition of culture and Leavis's presumption of an effective end to creative composition all aimed to socially anchor 'culture' quite completely. The tension between ideal and anchored meanings had arguably been lost.

To a significant degree, Williams's efforts had effectively restored these progressive Enlightenment dimensions by his immanent critique – as well as the very counter-selection – of his British 'culture and society tradition'. This recovery is most explicit in *The Long Revolution*'s discussion of the ideal conception of culture where Williams provides a major qualification of his 'localism':

> I find it very difficult, after the many comparative studies now on record, to identify the process of human perfection with the discovery of "absolute" values, as these have been ordinarily defined. I accept the criticism that these are normally an extension of a particular tradition or society. Yet, if we call the process, not human perfection, which implies a known ideal towards which we can move, but human evolution, to mean a process of general growth of man as a kind, we are able to recognize areas of fact which the other definitions might exclude. For it seems to me to be true that meanings and values, discovered in particular societies and by particular individuals, and kept alive by social inheritance and by embodiment in particular kinds of work, have proved to be universal in the sense that when they are learned, in any particular situation, they can contribute radically to the growth of man's powers to enrich his life, to regulate his society, and to control his environment. We are most aware of these elements in the form of particular techniques, in medicine, production, and communications, but it is clear not only that these depend on more purely intellectual disciplines... but also that these disciplines in themselves, together with certain basic ethical assumptions and certain major art forms, have proved similarly capable of being gathered into a general tradition which seems to represent, through many variations and conflicts, a line of common growth. It seems reasonable to speak of this tradition as a general human culture, while adding that it can only become active within particular societies, being shaped, as it does so, by more local and temporary systems. (*LR*, pp. 58–61)

While the relentlessly masculine formulations provide testimony of this text's 'pre-feminist' composition, they also testify to Williams's adoption of the then orthodox conception of 'man' as a species. Williams makes explicit his desire to keep his model of culture open beyond the localism that dominates the rest of the discussion. Crucially, he breaks with the anchoring of Arnold's ideal of culture-as-perfection-as-best, but can only do so by an appeal to evolution, so removing a 'known goal' but maintaining, nonetheless, a linear conception of development. Nor is there any explicit embrace here of the Enlightenment conception of culture or reason.

The short but densely argued 'On High and Popular Culture' (Williams, 1974a) is perhaps the most significant later text in this context. It stands as the only likely successor to Williams's comparable writings of the late 1950s and early 1960s. This makes its invisibility in the scholarship on Williams all the more remarkable.

Williams opens by distancing himself immediately from the existent opposition between the Arnoldian definition of culture as 'high culture' and its populist rejection in the name of 'popular culture'. The debate, he announces, is 'intolerably confused by failures of definitions' (1974a, pp. 13–14).

His reformulation follows the Herderian creation of a distinction between a universal process and its realization within specific cultures. Herder's challenge to the Enlightenment conception of culture occurred principally within his historical writings. There the progressive developmental sense of culture could be prioritized more easily over the 'anchored' achieved state. In particular, he foregrounds the necessary social and educational determinants of any developmental process of culture (cultivation). Herder argues that the Enlightenment foundation of this progress in an undifferentiatedly abstract and 'innate' conception of reason is insufficient and risks a Europocentric anchoring. He thus advocates the operative and implicitly comparative pluralization, *cultures*. His role in the fostering of the usages of *folk culture(s)* and *popular culture(s)* was thus pivotal; as was his influence on the emergence of the discipline of anthropology. Herder himself published a highly influential collection of folksongs.[61]

Both the philosophical and the folkloric interventions of Herder were part of a broader movement amongst European intellectuals towards a 'discovery of the people' (*volk*). The principal means of this 'discovery' was the notation of 'folk' material directly from the recollections of living people.[62] These practices of the European 'folk', usually understood as the peasantry, were regarded as inspirational by most Romantics. Their 'discovery' aided the formulation of a series of central critical tenets of Romanticism: the preference for 'primitivist' or 'exotic' artworks ('cultural primitivism'); the related hostility to the formal rules of neo-classicist composition and criticism, especially in poetry; the celebration of spontaneous creativity.

The last of these in particular implied a confidence in the capacities of the folk *themselves* as creative subjects. More commonly, however, they were regarded as at best semi-conscious bearers of an almost lost body of work. The rush amongst intellectuals to collect the folksongs and folktales was as much a process of conservation of tradition(s) as discovery. This is indicative of a further tendency amongst the Romantics: an historical retrospectivity in their cultural primitivism. The folk material in this context was a link with an idealized past.

For Herder's position was also constitutive of Romanticism's most consistent critical court of appeal that we have already met in Leavis, the *organic community*. The folk's directness, spontaneity and lack of pretension led many, including Herder, to tend to see them even as part *of* nature. But, again, the contemporary peasantry were regarded only as a suggestion of a lost age of complete unity between humans and nature. 'Organicist' metaphors, comparing social forms with such things as the fertility of the soil and vegetation, thus abound in Herder's and other Romantic writers' social commentaries.[63]

Significantly, across his assessments, Williams emphasizes the anti-Europocentric and anti-metaphysical implications of Herder's usage. He omits mention of the relationship between it and the development of consequent conservative-nationalist articulations of the pluralizing and relativizing subcategories, especially *folk culture*. The appeal to the example of a stable peasant culture was a key manoeuvre in conservative and reactionary arguments in the wake of the French Revolution: cultural stability could thus be used to foster nationalist resentment to Napoleonic occupation and even endorse the political order of the *ancien regime*. Herder himself, despite his overt populist sympathies, recoiled from the French Revolution.[64]

This conservative characterization is more typical of modern critico-philosophical accounts of Romanticism. Herder's initiative in these is seen as part of an expansion of a relativist and irrationalist assault on the Enlightenment's then recently consolidated achievements. The pluralization of 'culture' is seen to risk a slippage into a plurality of cultural evaluations. The undermining of the 'universalist' conception of culture as the summation of human self-development is seen to be a reversion to pre-Enlightenment religious and mystical beliefs. To undermine the universalist conception of culture was thus to undermine the universalist conception of reason.[65]

In 'On High and Popular Culture' Williams simply notes the significance of Herder's argument for the 'ordinary modern use of "culture" in anthropology' and then sets it aside. Likewise, the folkloric dimension is largely ignored.

Rather, it is the legacy of this contradictory dynamic for modern societies that concerns Williams most. His pluralized reformulation of culture's definition is thus: 'at once the general process of human development and the specific organizations of such development in different societies. It implies also both the whole way of life of a people and the practices and products of intellectual work and the arts' (1974a, p. 14). This evidently still has much in common with the passage from *The Long Revolution* (above) and, indeed the conjunction of meanings with which this chapter opened. But in 'development', Williams finally adopts the standard exegetical term for the Enlightenment conception of progress. 'Organization' is still present as well but, as we shall see, is about to receive tighter definition.

In a significant recent intervention in debates about the concept of culture, Robert Young has demonstrated the degree to which the Romantic translation of eighteenth-century Enlightenment arguments was also, however, inflected by nineteenth-century conceptions of European racial superiority. The Romantics' 'passion for ethnicity' could also be employed to argue for the 'permanent difference of national-racial types' (Young, 1995, p. 42). For

Young a kind of intellectual 'displacement' informs both the Enlightenment and Romantic projects: a process of projecting inner dissensions 'outwards into a racialized hierarchy of other cultures' (Young, 1995, p. 52).

It is clear that a key feature of the Enlightenment 'universal histories' was a staged model of human development. Young summarizes these as an eighteenth-century 'four-stage model' – Prehistory, Ancient, Medieval, Renaissance (to the then present) – which was succeeded by a nineteenth-century 'three-stage model' – Savagery, Barbarism, Civilization. Herder's break with this hierarchization is by no means complete. While he laid the ground for the plural formulation, cultures, he was still capable of assessing these cultures on the three-stage scale.[66]

In 'On High and Popular Culture' Williams's placement of Herder emphasizes the capacity of the category of culture (rather than civilization) to sidestep such dangers in the barbarism/civilization dichotomy.[67] Accordingly, his internationalized reference to 'high culture' reconstructs the critique of the selective tradition as a critique of imperialist and neo-colonialist relations between nations and societies:

> Thus whether within or between societies, respect for "high culture" in its purest and most abstract sense must find a critical rather than a justifying form of expression and action. (1974a, p. 14).

Williams does also provide a minimal 'skill-based' definition of high culture:

> Its most plausible use is to describe the great body of cultural skills and the great works which embody and represent them. There would be argument about which skills to include or exclude, but in common usage the skills of organized thought, writing, music, the visual arts and architecture would certainly be included. (1974a, p. 14)

What links these two accounts of high culture is Williams's view that 'high culture has no specific social structure' by which he means, initially, that high culture has no 'class belonging'. This is consistent with his explanation of 'culture is ordinary' and the 'common inheritance' formulation within the critique of Hoggart and *Culture and Society*. But this reference to social structure also starts to flesh out the ubiquitous category of 'organization' to which Williams continually refers in *The Long Revolution*. Instead of a social structure, high culture has a professional structure, that is, those who create new work with those cultural skills as well as those who select, maintain and disseminate the traditions. The critical form of 'expression and action' of high culture is necessitated by the pressure of social structures upon these professional structures, that is, pressures to turn high culture to a legitimating purpose.

Here at last Williams appears to have sketched a fuller answer to the challenge of Eliot's crude linkage of a dominant class and the cultural élite. Moreover, by also providing a theoretical recognition of the risk of high culture being turned to a legitimating purpose, Williams provides a fuller means of 'reconciling' culture as arts and learning and culture as 'whole way of life'. It is very clear that this legitimating role would not merely be the use of high culture as a means of displaying social status. Rather, such legitimation would be that of an existing social order, whether within a nation-state or within colonialist/imperialist relations between nation-states or upon 'whole ways of life'.

Moreover, the international commonality of these cultural professions makes possible the postulation of an international high culture that is very carefully defined within this critical mode:

> Between societies, when in any good faith the selective character of particular versions of high culture will quickly become obvious, we must explore the connections between these variations and the real historical and contemporary political and economic relationships, and, above all, to avoid the error of supposing that a selective version made by some temporarily dominant society is "universal" whereas the selective version of some temporarily dominated society is merely "local" or "traditional". The interaction between particular local selections and what can be conceived theoretically as a universal high culture must, for cultural as well as other reasons, take place in conditions of equality and mutual respect. This, of course, does not mean that what is sought is some bland consensus; there is much necessary opposition and conflict between variant cultural traditions, as well as honest recognition of alternatives. (1974a, p. 15)

Accordingly, '...there can be no simple contrast between "high culture" (universal) and "popular culture" (local)' (1974a, p. 15). Each informs the other without removing the distinction between them. All cultural institutions, even universities, are shaped and coloured to some degree by the 'popular culture' *of* a people. With that small emphasis, Williams insists on his familiar criterion of popular control regarding the latter category. Yet he offers no alternative terminology for what he calls a popular culture produced by 'commercial saturation'.[68]

Likewise, the process of communicative mutual recognition and cross-selectional diversity between cultures described above also reformulates a 'localist' version in the earlier work. Williams's distinctively non-uniform 'common culture' was to be based in his 'educated and participating democracy'. The cultures at stake include those non-reductively based in class as well as the 'common inheritance'.

Much of this 'long revolution' was premissed on the meeting of working-class institutional ideals embodied in democratic institutions with that 'common inheritance'. Williams's proposed minimal knowledge base for 'every educationally normal child', for example, included 'extensive practice in democratic procedures' as well as 'history and criticism' of the arts (*LR*, p. 175). His optimistic speculations about the common culture that would result from 'the coming to relative power and relative justice of your own people' today echoes the magnanimity of some recent leaders of post-authoritarian administrations. Williams speaks similarly of moving beyond 'the defensive practice of solidarity to the wider and more positive practice of neighbourhood' and even of 'achieving diversity without creating separation' (*C&S*, pp. 333–4).

As we saw, however, this modest optimism in a movement towards a participatory democracy came undone in the mid-1960s. It was not that Williams was naïvely unaware that powerful forces were developing and implementing a very different model of the future. Rather, his contingent judgement was that the balance of forces was such that the advocacy was worth the attempt. Even when that balance clearly shifted against his own (non-prescriptive) project, as it did increasingly during the remainder of his lifetime, his advocacy continued.

Of considerable relevance here is Williams's recognition in his mature engagements with the concept of culture that Marx's work could be placed within this broader post-Romantic Enlightenment frame.[69] Marx's identification of the contradictory form of the 'progressive' Enlightenment project enabled Williams to develop an account of the contradictory productive forces at stake in cultural modernity. Pivotally significant here is the role of what he came to call means of communication. An examination of Williams's journey towards this position via his engagement with 'received Marxist theory' is thus an appropriate next step in this account.

2
Cultural Materialism versus 'Received Marxist Theory'

2.1 Cultural materialism: a modest proposal

> It took me thirty years, in a very complex process, to move from that received Marxist theory (which in its most general form I began by accepting) through various transitional forms of theory and inquiry, to the position I now hold, which I define as "cultural materialism". The emphases of the transition – on the production (rather than only the reproduction) of meanings and values by specific social formations, on the centrality of language and communication as formative social forces, and on the complex interaction both of institutions and forms and of social relationships and formal conventions – may be defined, if anyone wishes, as "culturalism", and even the crude old (positivist) idealism/materialism dichotomy may be applied if it helps anyone. *What I would now claim to have reached, but necessarily by this route, is a theory of culture as a (social and material) productive process and of specific practices, of "arts", as social uses of material means of production (from language as material "practical consciousness" to the specific technologies of writing and forms of writing, through to mechanical and electronic communications systems).* I can only mention this here; it is spelled out more fully in *Marxism and Literature* and *New Sociology: Culture*.[1] What bears on this note is that what turned out to be, when developed, a materialist (but non-positivist) theory of language, of communication and of consciousness was assigned, along the way, to "idealism" just because, in received Marxist theory, these activities were *known* to be superstructural and dependent – so that any emphasis on their specific primacies (within the complex totality of other primary forms of the material social process, including those forms which had been abstracted as "labour" or "production") was known *a priori* to be "idealist". (1976c, pp. 88–9; *PMC*, p. 243; italicizations other than 'known' and 'a priori' added)

There is no definitive monograph called *Cultural Materialism* within Williams's corpus. Instead, the cultural materialism was modestly and sporadically announced in a series of publications from 1976 to 1980. The 'manifesto' passage above is the very first and most programmatic. In the following year's introduction to *Marxism and Literature* there is a similarly autobiographical announcement, and a briefer definition of the project as 'a theory of the specificities of material cultural and literary production within historical materialism' (*M&L*, p. 5).

The implications of this announcement for the sociology of culture are discussed in Chapters 3, 4 and 5. This chapter first examines Williams's critique of that orthodox 'received Marxist theory', especially as it manifested in discussions of Marx's own work. As with Chapter 1, this chapter has been shaped to some extent as a response to some important critiques of Williams. Significantly, the article from which the above passage is taken also functioned as Williams's point-by-point reply to Terry Eagleton's Althusserian critique of his work earlier the same year. The continuing influence of Eagleton's critique has been considerable, especially as it sets up some of the terms of Stuart Hall's later critiques.[2] These and related criticisms are discussed in this chapter.[3] The central question raised by these critiques is Williams's adherence or non-adherence to the 'classical Marxist' tenet of socio-economic *determination* of cultural forms. This is usually referred to as the problem of the 'base and superstructure' metaphor. As with the related critiques introduced in Chapter 1, I reject these as misplaced in their primary assumptions.

Instead, I elaborate the premisses of the position Williams advances in the above proposal. For it is evident that Williams resolves the arguably oxymoronic nomenclature, 'cultural materialism', by an appeal to a *Marxian paradigm of production.* He makes plain that the cultural materialism is not driven principally by an extended philosophical elaboration of a 'materialism' but, instead, by the application of this production paradigm to the field of culture.[4] This chapter introduces and develops this thesis.[5]

2.2 Back to Marx but beyond base and superstructure?

'Received Marxist theory' meant for Williams an orthodoxy that he had first met via his brief membership of the British Communist Party. In both *Culture and Society* and the 1976 essay cited above, Williams draws a clear distinction between a radical Romantic populist British tradition best represented by the work of William Morris on the one hand, and the derivative 'British Marxism' of the 1930s subordinated to the 'directive' Leninist conception of the vanguard party on the other.

Thus the 'Marxism' Williams unsympathetically reviews in *Culture and Society*'s chapter on 'Marxism and Culture' is the same local tradition which had failed him as an undergraduate at Cambridge during his own prior, 'orthodox' phase.[6] The consistent theme is the lack of theoretical resolution by the 1930s English Marxists of the influence of three sources: the available works of Marx, the appeals of the local Romantic tradition and Leninist vanguardism.

The work of William Morris, in particular, is seen to set a pattern of reconciliation between a relatively limited understanding of Marx and a populist radicalization of the Romantic critique of capitalism. Morris quite explicitly abandoned the clerisist strategy of cultural renewal for a declared alliance with the organized working class. Because of this achievement, it is Morris, of all those nineteenth-century writers examined in *Culture and Society*, who is 'the pivotal figure of the tradition' for Williams (*C&S*, p. 161).

'Culture is Ordinary' provides a succinct account of the limitations of 'received Marxist theory' and its expectation of the role of artists and culture:

> I did some writing while I was, for eighteen months, a member of the Communist Party, and I found out in trivial ways what other writers, here and in Europe, have found out more gravely: the practical consequences of this kind of theoretical error. In this respect, I saw the future, and it didn't work. The Marxist interpretation of culture can never be accepted while it retains, as it need not retain, this directive element, this insistence that if you honestly want socialism you must write, think, learn in certain prescribed ways. A culture is common meanings, the product of a whole people, and offered individual meanings, the product of a man's [*sic*] whole committed and personal social experience. It is stupid and arrogant to suppose that any of these meanings can in any way be prescribed; they are made by living, made and remade, in ways we cannot know in advance. To try to jump the future, to pretend in some way you *are* the future, is strictly insane. Prediction is another matter, an offered meaning, but the only thing we can say about culture in an England that has socialized its means of production is that all the channels of expression and communication should be cleared and open, so that the whole actual life, that we cannot know in advance, that we can know only in part even while it is being lived, may be brought to consciousness and meaning. (*ROH*, pp. 8–9)

A considered politico-theoretical point is present within this polemic. The (received) Marxist interpretation of culture is unacceptable because it contains this prescriptive view of cultural innovation in both the present and the promised socialism (as in orthodox Soviet 'socialist realism'). But it need not be so. Marx is clearly retrievable from 'received Marxist theory'. In effect, Williams deduces the possibility of what was later called a 'Western Marxist' position. Thus, in the place of 'received Marxist theory', he substitutes his own redefinition of the culture–class relation which is more fully developed in the critique of Hoggart examined in Chapter 1. From this an indication of his alternative, non-vanguardist, non-élitist, socialist future becomes possible in the final sentence above.

A similar position is argued in *Culture and Society*'s 'Marx on Culture' chapter. There the failings of Leninism are explicitly named as such, as in Lenin's conception of the directive role of the party towards artists. But Williams closes the chapter with a stringent critique of Lenin's delimitation of working-class consciousness unaided by the party to 'trade union consciousness'. The inability of 'received Marxist theory' to understand either artistic innovation or working-class consciousness remained an issue for Williams, as was the very linkage of the two within his own development of alternative models. For clearly the Leninist conception was entirely at odds with his own strategy (in that period) of drawing on working-class forms, in his immanent critiques, as a radical benchmark of the possible expansion of delimited conceptions of culture and democracy.

The questionable theoretical foundation of 'received Marxist theory' was always for Williams the legacy of reception of Marx's 'base and superstructure' metaphor. His favourite example of its vulgar use was the category of 'capitalist poetry' which came from the 1930s British Marxist, Christopher Caudwell.[7]

As Williams highlighted as early as *Culture and Society*, Marx used the metaphor on different occasions with significantly different emphases. But it was this famous passage in the summary text popularly known as 'The 1859 Preface' which became easily the definitive source:

> In the social production of their life, men enter into definite relations that are indispensable and independent of their will, relations of production which correspond to a definite stage of development of their material productive forces. The sum total of these relations of production constitutes the economic structure of society, the real foundation, on which rises a legal and political superstructure and to which correspond definite forms of social consciousness. The mode of production of material life conditions the social, political and intellectual life process in general. It is not the consciousness of men that determines their being but, on the contrary, their social being that determines their consciousness. At a certain stage of their development, the material productive forces of society come in conflict with the existing relations of production or – what is but a legal expression for the same thing – with the property relations within which they have been at work hitherto. From forms of development of the productive forces these relations turn into their fetters. Then begins an epoch of social revolution. With the change of the economic foundation the entire immense superstructure is more or less rapidly transformed. In considering such transformations a distinction should always be made between the material transformation of the economic conditions of

> production, which can be determined with the precision of natural science, and the legal, political, religious, aesthetic or philosophic – in short, ideological forms in which men become conscious of this conflict and fight it out. (Marx, 1958a, pp. 328–9)[8]

Much conventional critical wisdom would have it that Williams abandoned the priorities made clear by Marx in this passage. Hall, largely in agreement with Eagleton, believes Williams abandons the metaphor completely for a relativized societal model of 'indissoluble elements of a continuous socio-material process'.[9] This 'interactionist' perspective, Hall believes, is broadly consistent with that Williams first articulated about 'culture' in *The Long Revolution*. Hall thus aims to consolidate his interpretation of Williams's approach as 'culturalist'.[10]

This is easily the most serious misunderstanding of Williams's mature position that has occurred and moves in almost perfect parallel with the reception of his analysis of 'culture'.

Williams's reconstructive procedure with the metaphor and related texts is first announced in the most famous of the 'rapprochement' articles, the 1973 'Base and Superstructure in Marxist Cultural Theory', and continues throughout the mature project. It reaches a culmination in the neglected 1983 essay, 'Marx on Culture'.[11] Williams certainly does embrace the phrase, 'indissoluble elements of a continuous socio-material process', in his later work in order to assert his cultural materialist position. 'Indissolubility' is usually taken to mean insusceptibility to breakdown into smaller parts. The 'indissolubility' thesis is designed to set limits upon any theoretical *a priorism* that might subsume one 'element' into another, most obviously 'culture' into 'economy'. But plainly this thesis is a 'clearing operation' designed to maintain a theoretical space for Williams's own elaborated concepts, especially in his sociology of culture.

Eagleton, Hall and other commentators on Williams's relationship with Marxism (e.g. Márkus) understandably prioritize the 1973 essay. It is easily Williams's most trenchant critique of the metaphor but it is actually *atypical* in that it is entirely focussed on a critique of its *epochal* usages deriving from Marx's 'The 1859 Preface'. This epochal usage understands 'the base' to refer to the mode of production, the defining feature, for Marx, of an epoch (e.g. capitalism). But elsewhere Marx employs the metaphor so that the base also refers to more localized socio-economic determinants such as social classes or class fractions. These 'conjunctural' determinants may operate within a much smaller time frame than an epoch. In *Marxism and Literature* and 'Marx on Culture' Williams reasserts the interpretation he had adopted in *Culture and Society*, and which endorses Marx's 'other' usages of the base and superstructure metaphor. In doing so, Williams develops a position similar to Hall's, in that both rely heavily on Marx's *The Eighteenth Brumaire of Louis Bonaparte*.[12]

In order to avoid further confusing an issue so central to understanding Williams's cultural materialism, I will commence with a summary assertion that anticipates core components of my conclusion: on the specific issue of the viability of base and superstructure as a model of social determination in cultural analysis, Williams abandons the vulgar interpretation of its usage in 'The 1859 Preface' for Marx's complex operative usage in the case of the 'intellectual representative' in *The Eighteenth Brumaire of Louis Bonaparte*. He places definite limits on the applicability of even this complex operative usage. He never sees it as a 'universal' paradigm of cultural analysis but it does play a pivotal role in his mature project. However, clarifying the role of this metaphor does not complete the elucidation of the cultural materialism nor its relationship with Williams's sociology of culture.

2.3 'The Brumaire solution' and the attractions of homological analysis

Even as it stands in 'The 1859 Preface' version, the metaphor is far more conceptually sophisticated and dynamic than its vulgar practitioners usually suggest. A limited defence of the metaphor can thus be mounted.[13] The 'material productive forces' are composed of labour and technical means of production, the relations of production being fundamentally defined by ownership or non-ownership of such means of production. 'The base' is constituted by the productive forces and relations, which together define the 'epochal' *mode* of production (e.g. capitalism).

The pressure of increased 'productivity' upon the superstructure comes primarily through the revolutionization of technical means (which capitalism especially has achieved). The productive forces thus stand in a dialectically *contradictory* relation with the relatively stable productive relations ('fetters'). The historical dynamic so induced places pressure on the superstructure's 'epochal' 'legal and political' forms. A 'social revolution' (involving a change in productive relations) results in which the superstructure is 'more or less rapidly' transformed. The paradigmatic 'epochal' events for Marx are the French revolutions from 1789 to 1871.[14]

Vulgar Marxism tends to reduce this complex dynamic to a static reflective relationship between a poorly defined 'economic' level and superstructural forms – derived from it – hence Caudwell's 'capitalist poetry'. Marx's 'precision of natural science' became all too easily the unbending 'iron laws of history'. These were tendencies Williams opposed and contrasted with Marx's actual texts and practice, but his central difficulties are already present in the passage above. The Caudwell case also exemplifies perhaps the key feature of much vulgar reductivism, the misapplication of Marx's *epochal* understanding of the base to the minutiae of short-term *conjunctural* cultural changes in 'non-epochal' superstructural forms.

Employing his 'contradictory passages' technique of immanent critique, Williams pointed to the alternative usage of the base and superstructure metaphor as early as *Culture and Society*. Partly as a tactic to reveal the limited understanding of Marx by 1930s English Marxists, he pits this passage from *The Brumaire* against that (above) from 'The 1859 Preface':

> Upon the several different forms of property, upon the social conditions of existence, a whole superstructure is reared of various and peculiarly shaped feelings, illusions, habits of thought and conceptions of life.
>
> The whole class produces and shapes these out of its material foundation and out of the corresponding social conditions. The individual unit to whom they flow through tradition and education may fancy that they constitute the true reasons for and premises of his conduct. (Marx, 1951, p. 62)[15]

Williams's juxtaposition of the two uses of the metaphor undercuts the legitimacy of reducing all 'superstructural' phenomena to epiphenomena of a mode of production. *The Brumaire* was written as an analysis of the defeat of the 1848 Revolution in France and the subsequent coup by Louis Bonaparte in 1851. It was one of the first occasions on which Marx brought his historical materialist concepts to bear in what Williams would call 'actual historical' analysis. It thus provides considerable evidence for the case that 'vulgar Marxism' was largely a product of the *unmediated* application of the 'epochal' meaning of 'the base' provided in 'The 1859 Preface'.

One of *The Brumaire*'s central tasks is to provide an account of how the legal and political superstructure, a state, does 'more or less rapidly transform' in response to the determinant pressure of changes in 'the base'. In order to do this, Marx needs to employ his own localized form of the metaphor to demonstrate the gestation of political ideas from 'the social conditions of life' of a social class.

The above is only one of many conceptual additions to the apparent simplicity of the usage in 'The 1859 Preface' that Marx makes in this more developed application of the base/superstructure model.[16] Social classes are examined within their self-organizing units of *fractions* and *strata*. These are seen to form alliances that find representation in the political superstructure as *blocs*. This representation is not conceived as the maximal liberal-democratic one of self-conscious personal delegate. Rather, the representation requires no such individual conduit or mirror reflection between, say, bourgeois party and bourgeois class. Quite unlikely social forces may come into alliance and find even more unlikely means of representation in the 'political theatre', as Marx often describes it, of the superstructure.

The real significance of this formulation is the way in which Marx sees this 'unlikely' representation working. As Hall usefully suggests, such representation can be seen as a process of '*re-presentation*' (Hall, 1977a, p. 44).

Representation is conceived here as a process of determination by a reproduction (re-presentation) of the same 'pressures and limits' on the representative as on those represented.[17] This discussion of the representative role of the Social Democratic Party is the key passage from *The Brumaire*:

> What makes them representatives of the petty bourgeoisie (though "according to their education and individual position they may be as far apart as heaven and earth") is the fact that in their minds they do not get beyond the limits which the latter do not get beyond in life, that they are consequently driven, theoretically, to the same problems and solutions to which material interest and social position drive the latter practically. This is, in general, the relationship between the *political* and *literary representatives* of a class and the class they represent. (Marx cited in *WICTS*, pp. 223–4; cf. Marx, 1958b, p. 250)[18]

Williams immediately comments on this passage:

> This can be taken too simply, but it is the source of the important modern Marxist conception of *homology*, or formal correspondence, between certain kinds of art and thought and the social relations within which they are shaped. This conception can reveal determining relations at a quite different level from the bare proposition that "ideas are nothing more than the ideal expression of the dominant material relationships"; among other reasons is the fact that something more than reflection or representation is then often in question, and art and ideas can be seen as structurally formed, but then also actively formed, in their own terms, within a general social order and its complex internal relations. (*WICTS*, p. 224)

The last is indeed the crucial general point *The Brumaire* exemplifies for Williams – the recognition of the immanent 'active' development of cultural forms 'in their own terms'. It is a central condition of his theoretical re-affiliation with Marxism.

The Brumaire model does indeed provide sufficient autonomy to cultural forms such that their alignment with specific social forces is seen as a determinate product of their 'homologous' but convergently corresponding role. They are not reflectively 'provided' for just that purpose of alignment, nor is that alignment 'guaranteed'.

The fuller significance of Williams's subsequent endorsement of *The Brumaire* would appear to be that it is the chief of Marx's texts found to meet these criteria. Perhaps it always did for Williams but it was not until 1983 that he was either able or prepared to make this so explicit.[19]

These assertions sit uneasily with the received critical wisdom that Williams was hostile to the base and superstructure metaphor.[20] Indeed, the significance of the above endorsement of *The Brumaire* for 'positioning'

Williams theoretically is difficult to overestimate. It is impossible to reconcile with the view that he rejected the base and superstructure metaphor out of hand. Here for Williams, in a passage to which he had long had access, was an account of the determination of the 'superstructural' which also acknowledged the formation of signifying practices 'in their own terms', especially the terms made available by (national) traditions.

Any doubts are completely removed by examination of Williams's analyses of the Bloomsbury group.[21] To anticipate the discussion in later chapters, although *The Brumaire* is not cited, this analysis reads as if it were a direct application of the 'intellectual representative' passage from it above. Williams treats the Bloomsbury group precisely as if it were a contending representative force on the 1848 French political stage.

We can reach a similar assessment by a different route by considering Williams's historical semantic discussion of the concept of 'determination'. His analysis has delivered an influential redefinition of 'determination' as the setting of 'pressures and limits'. This also is entirely compatible with the discussion of homologous 'limits' of the 'class representative' in the passage from *The Brumaire*.

However, Williams's discussion of this dimension of 'pressures' goes further. Williams's initial comments would appear to support the criticism that he becomes prone to a 'voluntarist subjectivism':

> For in practice determination is never only the setting of limits; it is also the exertion of pressures. As it happens this is also a sense of "determine" in English: to determine or be determined to do something is an act of will and purpose. (*M&L*, p. 87)

This certainly suggests a voluntarist subjectivism that overemphasizes human agency.[22] It might be asked perhaps whether Williams means then that 'heroic individuals' are always so capable of challenging the 'limits' of determinations? However, Williams's immediate elaboration of his comment moves in a quite different direction:

> In a whole social process, these positive determinations, which may be experienced individually but which are always social acts, indeed specific social formations, have very complex relations with the negative determinations that are experienced as limits. For they are by no means only pressures against the limits, though these are crucially important. They are at least as often pressures derived from the formation and momentum of a given social mode: in effect a compulsion to act in ways that maintain and renew it. They are also, and vitally, pressures exerted by new formations, with their as yet unrealized intentions and demands. "Society" is then never only the "dead husk" which limits social and individual fulfilment. It is always also a constitutive process with very powerful

> pressures which are both expressed in political, economic and cultural formations and, to take the full weight of "constitutive", are internalized and become "individual wills". Determinations of this whole kind – a complex and interrelated process of limits and pressures – is in the whole social process itself and nowhere else: not in an abstracted "mode of production" nor in an abstracted "psychology". (*M&L*, p. 87)

It is important to note that 'formation' is the category the later Williams uses to analyze groupings of intellectuals, especially artists. Indeed, when the passage is re-read in this context, it becomes obvious that Williams sees intellectual formations as the primary means of mediation between social determinations and cultural 'production'.[23] Also evident here is the granting of a privileged position to innovative intellectual formations which can be seen as consistent with his conception of 'structure of feeling'.[24]

This brings us to the issues related to Williams's use of the production paradigm and implicates the related category of 'cultural productive force'.

2.4 Enter 'cultural production'

Williams's (re)embrace of '*The Brumaire* solution' occurs at the end of 'Marx on Culture'. It is less the culmination of his argument than a consequence of his somewhat exhaustive assessment of certain ambiguous formulations within Marx – especially in *The German Ideology* (1845) – that contributed to the base and superstructure metaphor's legacy for 'received Marxist theory'. Many of the texts of Marx that Williams draws on there and in *Marxism and Literature* were unavailable to him at the time of writing *Culture and Society*. More than in any other text, it is in 'Marx on Culture' that we can see Williams's elaboration of the premise of his cultural materialism as a paradigm of production.[25]

Williams's assessment stresses that polemicism is a consistent feature of Marx's delineations of his historical materialist project. This polemic was developed in opposition to the 'heaven to earth' causalities assumed within the titular 'German Ideology' of nineteenth-century Idealism. As we saw in Chapter 1, Williams drew attention to the absence of any 'material of the process' in proposals for the achievement of the ideal of human perfection within British clerisism. Likewise, but with a more scathing polemic, most of Marx's pronouncements about German Idealism are directed at this same absence of consideration of socio-material determinants and presuppositions.

There is, for instance, this famous passage that forms part of the outline of a materialist conception of history from *The German Ideology* which also strongly resembles the close of the passage from 'The 1859 Preface':

> In total contrast to German philosophy, which descends from heaven to earth, we here ascend from earth to heaven. That is to say, we do not set

> out from what men say, imagine, or conceive, nor from what has been said, thought, imagined or conceived of men, in order to arrive at men in the flesh. [We begin with real active men ... etc.] (Marx cited in *WICTS*, p. 203)[26]

Williams provides this commentary on the passage which is one of the best summations of his criticisms:

> As a statement of philosophical presupposition this is clear and admirable. It is wholly consistent, in its general emphasis, with the argument that we must begin any inquiry into human development and human activities from actual human beings in their actual conditions. But then rather more than this is actually said. The rhetorical reversal of metaphysical thought, in the proposal to "ascend from earth to heaven", has the extraordinary literal effect, if we are reading it closely, of shifting "what men say, imagine or conceive" and "what has been said, thought, imagined or conceived of men" from earth to ... heaven! Of course Marx did not literally believe this. It is a by-product of that particular polemical rhetoric. Yet a more serious question underlies the idiosyncrasy of the particular formulation.
>
> In this way of seeing the problem, and in fact against other emphases by Marx elsewhere, there is a real danger of separating human thought, imagination and concepts from "men's material life-process", and indeed of separating human consciousness from "real, active men". Taken crudely and literally, as indeed it has sometimes been taken, this is, ironically, a familiar position of bourgeois philistinism, of the kind satirized by Brecht as "eats first, morals after", or more seriously of the kind now regularly propagated by apologists of capitalism, in the argument that we must first "create wealth" and then, on the proceeds, "improve the quality of life".
>
> Marx's central emphasis was so much on the necessary totality of human activity that any reduction of this kind has to be firmly rejected. (*WICTS*, p. 203)

The serious theoretical risk Williams perceives, then, is to subordinate culture in the name of the historical materialist cause. As with earth and heaven, so with base and superstructure. A precondition of the reflectionist understanding of the determinacy described in the metaphor, is to reduce culture to an 'immaterial' phenomenon in contrast with real 'activity'.

For Williams, there is a 'more adequate conception' of human activity within Marx's understanding of human labour, best exemplified in his presuppositions about human labour in the chapter on the labour process in *Capital*:

> We pre-suppose labour in a form that stamps it as exclusively human. A spider conducts operations that resemble those of a weaver, and a bee puts to shame many an architect in the construction of her cells. But what distinguishes the worst architect from the best of bees is this, that the architect raises his structure in imagination before he erects it in reality.[27] At the end of every labour-process, we get a result that already existed in the imagination of the labourer at its commencement. He not only effects a change of form in the material on which he works, but he also realises a purpose of his own that gives the law to his modus operandi, and to which he must subordinate his will. (Marx, 1974, p. 174)

Williams thus comments:

> This convincing account of the specifically human character of work includes... not only the foreseeing concept of what is being made but ideally integrated concepts of how and why it is being made. This is intended to enforce Marx's conception of what is truly human in labour, and thus to provide a standard from which it is reasonable to describe certain forms of human work... as degraded or sub-human, in no hyperbolic sense.... Thus "real active men", in all their activities, are full of consciousness, foresight, concepts of how and why, or to the degree that they are not[,] have been reduced from this fully human status...
>
> But then it remains very strange that in the early writings, in which he wrote most directly of what we now call "cultural" activities, Marx worked with so vulnerable a definition of consciousness. It can of course be argued that what he then had mainly in mind was not the integrated consciousness of necessary human labour and genuine production, but what he and others could see as the phantasmagoria of religious and metaphysical speculation or the self-justifying systems of law, politics and economic theory which ratified oppression, privilege and exploitation. (*WICTS*, p. 204)

In the final sentence above, Williams acknowledges the legitimacy of Marx's prioritization of 'material activity' over consciousness as a critique of legitimating ideologies. But this is an exceptional instance. His subsequent argument in 'Marx on Culture' attempts to reconcile the disjunction between the 'integrated consciousness' of Marx's conception of human labour, and the defensible but ambiguous prioritization in the earlier writings. Whatever the defence of this prioritization, Williams initially notes that its chief risk is 'a very puzzling combination of historical and categorical argument'. He briefly considers but sets aside another line of defence – that certain human needs might be prioritized as more 'basic' – because it is an insufficiently historicized defence.

The argument Williams respects most is that based on Marx's own historicization of the 'breaking up' of that norm of an integrated human consciousness: the division between 'mental' and 'material' labour. Clearly this, rather than a hierarchy of human needs, provides one of Marx's primary distinctions in 'both' uses of the base and superstructure metaphor in the 1850s. Here too, however, Williams calls Marx to account with one of his alternative formulations: 'The organization of the division of labour varies according to the instruments of labour available.'[28] Yet, as we have seen, instruments of labour are clearly left in 'the base' in 'The 1859 Preface' as a component of the forces of production.

Marx so appears to have overlooked the specificity of the instruments of labour of 'mental labour'. This contradiction in Marx enables Williams to make this explicit claim:

> But then this same point is highly relevant to the actual processes of "mental" labour. Even if we retain, at this point, his categorical distinction between "material" and "mental" labour (overriding... the diverse social and historical conditions within which this distinction is variably practised and theorised), it soon becomes clear, from historical evidence, that *the productive forces of "mental labour" have, in themselves, an inescapable material and thus social history*. (*WICTS*, p. 211; emphasis added)

This is Williams's most explicit declaration of the conception of *cultural productive forces* that is quite crucial to his mature sociology of culture.[29] It was developed, as we shall see in Chapter 6, from his reflections on 'the media' as examples of 'means of cultural production'. Williams quickly adds his insistence here, however, that the division of labour in question 'cannot be reduced to a history of technical means alone'.

Williams's final assessment of Marx shares much with György Márkus's assessment of the role of 'production' in Marx's work. Márkus argues that there is a production paradigm, grounded in that same normative conception of human labour, throughout Marx's work. Marx sees labour as containing a conscious objectivating component. Acts of labour demonstrate a human capacity to design objects consciously in response to determinate needs. Crucially, however, this is not the postulation of an ontological human 'essence' which mysteriously manifests itself in acts of production.[30] Rather, labour-as-objectification provides the rationale for a paradigm of production in which productive activity is recognized as both physical and mental work. Two types of products are so produced, 'material' and 'cultural' (Márkus, 1986, p. 43).

As we shall see, on this last distinction, Williams and Márkus disagree. While Williams does adopt the terminology of 'objectivation' irregularly, his preference is clearly for 'cultural duplication' of the 'basal' categories of the base and superstructure metaphor as in Table 2.1. Thus Williams joins,

Table 2.1 Williams's cultural duplication of 'the base'

	Forces of production, i.e. labour plus these varying means of production	**Relations of production**
General	Means of 'general' production	Capital/labour (initially)
Cultural	Means of cultural production including means of communication	Relations of cultural production including 'formations'

apparently unwittingly, a considerable legacy of Marxian attempts to 'extend' the paradigm of production to culture and cultural analysis.[31]

Williams's 'cultural productive forces' would appear, at first glance, to 'collapse' the two tiers of determining base and determined superstructure into one. Indeed, it appears to raze the house Marx metaphorically built in 'The 1859 Preface'. But although some of Williams's critics insist that such must be the case, it does not follow, especially in his analytic practice. The metaphor may be under stress, but Williams's fundamental conceptual distinction between cultural productive forces and social/general productive forces is maintained. For Williams, they have been rendered less vulnerable to *a prioristic* reification.

Thus Williams does insist that his expansive categorical shift demonstrates that even vulgar materialists were 'not materialist enough', and that the superstructure component of the metaphor was an 'evasion' from the necessary materiality of cultural practices (*M&L*, pp. 91–2). Likewise, he acknowledges the apparent theoretical risks in the removal of the metaphor's determinant premiss:

> Yet the difficulty is that if we reject the idea of a "self-subsistent world" of productive (industrial) forces, and describe productive forces as all and any activities in the social process as a whole, we have made a necessary critique but, at least in the first instance, lost edge and specificity. (*M&L*, p. 93)

But this is Williams's rationale *for* 'cultural productive forces'. On the specific question of social determination only a revision of the metaphor's range of application is necessitated.

For we still have the original categories of the metaphor, as well as Williams's culturally specified versions. The determinant role of 'the base' over a formal superstructure in any general sense is thus supplanted by the relation between what have now been constituted as two sets of productive forces and relations: cultural and 'social' (or 'general'). The key to the relationship between these elements at any determinate moment is approached through their common but differentiated processual dynamics of *reproduction*: that is, social reproduction and cultural reproduction.

For Williams it is a matter of determinate conjunctural analysis, whether the relation between social and cultural reproduction is one of correspondent homology (including '*The Brumaire* solution'), 'asymmetry', or 'symmetry'.[32]

'*The Brumaire* solution' itself emerges as a further elaboration of Williams's efforts to render Marx's 'categorical' mental/manual division of labour in a more historically sensitive form. As we shall see in the next chapter, this project can be traced to his reception of the work of Goldmann and Gramsci. Gramsci too adopted *The Brumaire* model of analysis of a contemporary political conjuncture as a central premiss of his conception of hegemony. *The Brumaire* is a regular point of reference in Gramsci's prison notebooks entitled 'The Modern Prince' and 'State and Civil Society'. There, explicit and implicit comparisons between the cases of 1848–51 France and the rise of Italian fascism abound.[33]

2.5 Problems of 'cultural production': Márkus's critique

György Márkus is the most careful and sympathetic of the few commentators on Williams's later work, but he also argues that there are fundamental contradictions within Williams's deployment of the production paradigm.

Márkus argues that Williams faces the same conceptual difficulties as other practitioners of the production paradigm in cultural analysis, notably Adorno and Benjamin. That is, the very adoption of the paradigm of production/labour as 'material production' fails to address, indeed arguably renders conceptually impossible, the specificity of cultural objects as primarily bearers of cultural meanings.[34]

We saw earlier that the production paradigm in Marx recognizes not an ontological essence but the dual role of physical labour and 'designing' mental labour as two types of productive activity which generate correspondingly different types of objectivation. However, for Marx, cultural objectivations are not then 'outside' the realm of material production. Both forms of product are indeed the result of the bringing together of both forms of work.[35] The objectivating dimension of 'material' labour is precisely Marx's means of acknowledging this intellectual component in products with definite use-values. Cultural objectivations, self-evidently, are principally the products of intellectual labour, 'meaning-complexes embodied in some material form' (Márkus, 1990, p. 100).

To 'reintroduce' the production paradigm to that of cultural objectivation is thus potentially tautological. Indeed the danger of reductivism could return in a new form where 'meaning-complexes' are reduced to 'material form'.

For Márkus objectivation and materialization are the two key *but distinct* features of the Marxian production paradigm. *Objectivation*, as we have seen, is the human process of rendering 'human needs and abilities' into the object-form of material products in order to perform a specific use. This constitutes their 'material content'. *Materialization* refers to the simultaneous embodiment within those same products of definite social relations, their social 'forms'. The latter derive from the social conventions of 'proper' modes of consumption/appropriation of these objects.[36]

This dual characterization requires a process of reproduction as renewal both of objects' sheer physicality and of their social rules of use. Both provoke the larger question of societal reproduction.[37]

Consistent with this position, for Márkus, specifically cultural objects are further required to meet the criterion of novelty within the culture of modernity. They are exceptional objectivations, *unique* works of 'creativity'. For Marx they are thus 'ideal objectivations'. Rather than producers of *artworks*, artists are creative authors. Márkus finds then that attempts to apply the production paradigm to culture – that is, the very conceptualizations of 'cultural production' – fail especially at this level of reproduction. For while it is a defining feature of objects of utility that they require replacement (reproduction), often with duplicates, cultural objects as such do not. Instead, appropriate consumption by an appropriately competent public is the principal condition of a cultural object's 'immaterial' reproduction-as-survival. That is, the 'material form' of a cultural object is merely the bearer of a cultural meaning which is the appropriated/consumed component of the object. As such, it cannot 'materially' be destroyed or 'wear out'. Rather, its reproduction-as-survival as a cultural object is dependent on its continued 'cultural' consumption. Thus while objects of utility require regular reproduction by 'material' replacement, cultural objects require reproduction by regular cultural consumption. These, in effect, are the preconditions of an autonomous cultural sphere.[38]

Márkus so further underpins his view that 'cultural production' is a potentially tautological postulation. This problem manifests most obviously as an issue of cultural consumption. The tautology effectively destroys the distinction between cultural consumption and the consumption of other objects. Márkus thus says of attempts such as Williams's to employ a production paradigm:

> their emphasis, in my view, falls predominantly and one-sidedly, upon those social institutions which *pertain to* the sphere of culture, ensuring its integration into the total process of social reproduction, and not on the social relations *constituting* the realm of culture as such. (Márkus, 1990, p. 99)

Williams anticipates this line of criticism to some extent in 'Marx on Culture'. He even provides an appropriate citation from Marx which questions the very categorization of art as the objectivation of 'unique' labour. In his critique of Stirner in *The German Ideology*, Marx responds in the following manner to Stirner's attempted exemplification of such uniqueness in his remark that 'no-one can do Raphael's work for him':

> [He] imagines that Raphael produced his pictures independently of the division of labour that existed in Rome at the time. If he were to compare Raphael with Leonardo da Vinci and Titian, he would know how greatly Raphael's works of art depended on the flourishing of Rome at that time, which occurred under Florentine influence, while the works of Leonardo depended on the state of things in Florence, and the works of Titian, at a later period, depended on the totally different development of Venice. Raphael as much as any other artist was determined by the technical advances in art made before him, by the organization of society and the division of labour [in his locality and, finally, by the division of labour] in all the countries with which his locality had intercourse. Whether an individual like Raphael succeeds in developing his talent depends wholly on demand, which in turn depends on the division of labour and the conditions of human culture resulting from it. (Marx and Engels, 1977, p. 189)[39]

As Williams notes, Marx's observations on 'social environment' are commonplace in a present day 'identifiable "sociological" position'. However, the more interesting dimension here for Williams is Marx's underdeveloped comment on 'technical advances in art'. There is then an 'evident gap' in Marx's position 'between a briefly mentioned technical dimension and a general environment':

> it is in that gap, in that area of actual intersections between a material process, general social conditions, and the unmentioned assumptions about the purposes and content of art within those conditions, that the decisive question about the art itself are to be found. By including the specific social and historical conditions Marx has usefully broadened the scope of the inquiry, but has not then made it. (*WICTS*, p. 216)

It is not difficult to discern that Williams is here once again making a case for the introduction of another specialized usage of the category of productive force, the cultural productive force of technical advances within art. Indeed he quickly makes a similar case for social relations of cultural production and, in an echo of the cultural materialist 'manifesto' that heads this chapter, directs the reader to *The Sociology of Culture* for more detail.[40]

But does this position adequately 'answer' Márkus's critique? I would suggest it does by questioning, as does Marx's critique of Stirner, the adequacy of the criterion of uniqueness in the definition of art within cultural modernity. For Williams, as Márkus approvingly acknowledges, such categorical criteria must be sufficiently historicized before any such claims can be made. Moreover, much of *The Sociology of Culture* is, as we shall see, devoted to delineating the social constitution of Márkus's 'realm of culture as such'.

Yet Márkus also immanently criticizes Williams for his inconsistency in his application of this radical historicism within his own position. Williams also demonstrates for Márkus the problems inherent in attempts to reconcile this historicizing impetus with the specificity of cultural forms and, secondly, with cultural traditions which transcend 'epochs'. The task announced but never undertaken by Marx in the case of cultural forms is to provide reconstructions of their social genesis. In the case of cultural traditions which transcend 'epochs', there is Marx's well known and justifiably much-disparaged thesis in 'The 1857 Introduction' that the continuing attraction of Greek art and epic is due to their bearing the 'eternal charms' of the 'historical childhood of humanity' (Marx, 1973b, pp. 110–11).

Márkus finds that in much twentieth-century Marxian cultural theory these two issues have tended to become conflated as the same problem. Moreover, a recurrent type of solution emerges which owes much to Marx's 'eternal charms' thesis: a tendency to appeal to a dehistoricized conceptualization of 'maturing' cultural practices grounded in an anthropological conception of 'fundamental' genre categories.[41]

In a section of *The Sociology of Culture* (discussed in detail in Chapter 4), Williams appears to advocate just such permanent properties. He posits *modes* as 'deep forms' which can be related 'more to the sociology of our species, at a certain level of cultural development, than to the specific sociology of a given society at a certain place and time' (*SOC*, p. 150). Does not this conception of mode confirm Márkus's charge of a retreat into 'fundamental' genre categories by Williams?

But Williams immediately adds the following codicil:

> Yet these markedly longer phases and rhythms – these deepest forms – can no more be abstracted from general social development than they can be reduced to merely local conditions. (*SOC*, pp. 150–1)

Reduction, whether 'sociological' or anthropological, is clearly not an option for Williams here. While Márkus's critique turns on the charge that cultural forms and enduring cultural traditions become conceptually conflated, Williams, in fact, clearly separates the two issues. Traditions are not conceived as necessarily tied to modes or even genres or types.[42] Rather, traditions are located within an historical (i.e. conjunctural) rather than epochal mode of analysis. *Selective traditions* are products of social institutions of cultural reproduction, especially education and informal intellectual formations.[43]

Likewise, the constitution of a dominant cultural tradition is a key moment in the establishment of hegemonic rule within a specific social order (of a nation state).[44] Even more than his famous distinction between dominant, residual and emergent, this very recognition of tradition as a component of hegemonic practice could be seen as Williams's most significant contribution to the reception of the Gramscian model.

However, as we shall see in Chapter 3, Williams also shares with Adorno an understanding of cultural forms, especially modes, as cultural productive forces. While the reference to the 'sociology of our species' in the passage Márkus cites is significantly 'fundamental', the simultaneous invocation of 'a certain level of cultural development' provides the key link with Williams's historicism. Such cultural development is closely tied to the available 'technical' means of cultural production. Those available means – and attendant divisions of labour – are seen to facilitate 'disembedded' cultural practices.

Indeed, in 'Marx on Culture' Marx's 'eternal charms' passage provides Williams with a counterpoint case study to Marx's critique of Stirner's 'uniqueness'. Williams does not attribute Marx's 'extraordinary proposition' to a chosen dehistoricization but rather to his *reluctance* 'to apply the idea of material progress to the history of art' in this case. Unlike the case of Stirner on Raphael, 'his attachment to early Greek art was much too strong for that' (*WICTS*, p. 218). Clearly, Williams would radically historicize not only the issue at hand, but Marx's very assumption of the 'eternal charms'. So he moves on to this remarkable reflection on the historicization of reception and 'judgement' which provides his answer to the problem posed by Márkus of enduring cultural forms:

> Moreover, in the case of art, where simple physical consumption is not in question, no work is in any fully practical sense produced until it is also received. The social and material conditions of the original production are indeed stable: the material object (painting, sculpture) or the material notations (music, writing) are there, if they survive, once [and] for all. Yet until a further (and in practice variable) social and material process occurs, necessarily including its own conditions and expectations, the objects and the notations are not fully available for response. Often the varying conditions and expectations of response actually alter the object or the notation as it is *then* perceived and valued. Yet there are also some important continuities, which in Marxist terms do not relate to some unchanging pre-given human nature, nor to notions of the "childhood" or "maturity" of humanity, but to a range of human faculties, resources and potentials – some of the most important based in a relatively unchanged human biological constitution; others in persistent experiences of love and parentage and death, qualified but always present in all social conditions; others again in the facts of human presence in a physical world – with which certain works connect, in active and powerful ways, often apparently beyond the limited fixed ideas of any particular society and time. (*WICTS*, p. 220)

The introduction of a distinction between the different 'material conditions of the original production' – that is, notations and works – becomes quite pivotal in the mature project. It first appears in Williams's ambiguous formulations in the 1973 'Base and Superstructure' essay. The key section is the somewhat enigmatic conclusion entitled 'Objects and Practices'.[45] Márkus is drawn particularly to Williams's closing declaration there that 'we should look not for the components of a product but the conditions of a practice' (*PMC*, p. 48). He interprets this as the first sign of Williams's *de facto* embrace of the production paradigm and his 'decisive farewell' to the base and superstructure metaphor (Márkus, 1994a, p. 435). I have already argued that Williams does not abandon the base and superstructure metaphor completely and indeed actively re-embraced *The Brumaire* version in 'Marx on Culture'. With that matter put aside, it can be seen that in building his case summarized in the phrase Márkus cites, Williams tries to recognize the same points Márkus makes in his criticism of Williams's and Adorno's 'tautological' use of the production paradigm in relation to the constitution of 'culture as such'. However, this convergence of views comes *at the expense of Williams's rejection of a generic adoption of the categories of object/objectivation*:

> the true crisis in cultural theory, in our own time, is between this view of the work of art as an object and the alternative view of art as a practice. Of course it is at once argued that the work of art *is* an object: that various works have survived from the past, particular sculptures, particular paintings, particular buildings, and these are objects. This is of course true, but the same way of thinking is applied to works which have no such singular existence. There is no *Hamlet*, no *Brothers Karamazov*, no *Wuthering Heights*, in the sense that there is a particular great painting. There is no *Fifth Symphony*, there is no work in the whole area of music and dance and performance, which is an object in any way comparable to those works in the visual arts which have survived. And yet the habit of treating all such works as objects has persisted because this is a basic theoretical and practical presupposition. But in literature (especially drama), in music and in a very wide area of the performing arts, what we permanently have are not objects but *notations*. These notations have then to be interpreted in an active way, according to the particular conventions. But indeed this is true over an even wider field. The relationship between the making of a work of art and its reception is always active, and subject to conventions, which in themselves are forms of (changing) social organization and relationship, and this is radically different from the production and consumption of an object. It is indeed an activity and a practice, and in its accessible forms, although it may in some arts have the character of a singular object, it is still only accessible through active perception and interpretation. This makes the case of notation, in arts like drama and literature and music, only a special case of much wider truth. What this can show us here about the practice of analysis is that we have to break from the common procedure of isolating the object and then discovering its components. On the contrary we have to discover the nature of a practice and then its conditions. (*PMC*, p. 47)

The introduction of the category of *notation* is thus also quite crucial to the replacement of 'art object' by 'practice'. Rather than 'ideal objectivation', Williams reconceptualizes art as a practice constituted by *either* objects or notations – such as that in the paradigmatic example of the script of an enacted dramatic performance.[46] Indeed he regards the conceptualization of notational art as 'object' or 'text' as indicative of a 'consumption' orientation related to norms of taste rather than the prospects of further cultural production (*PMC*, p. 46).

The objectification/objectivation component of the Marxian production paradigm is thus less available to Williams. Unlike Márkus, he employs 'objectification' only rarely and then in a more narrow sense:

> Writers, in ways which we must examine and distinguish, handled material notations on paper.... It is only when the working process and its results are seen or interpreted in the degraded forms of *material commodity production* that the significant protest – the denial of materiality by these necessary workers with material – is made and projected into abstracted "higher" or "spiritual" forms. The protest is understandable but these "higher" forms of production, embodying many of the most intense and most significant forms of human experience, are more clearly understood when they are recognized as *specific objectifications*, in relatively durable material organizations, of what are otherwise the least durable though often the most powerful and affective human moments. The inescapable materiality of works of art is then the irreplaceable materialization of kinds of experience, including experience of the production of objects, which, from our deepest sociality, go beyond not only the production of commodities but also our ordinary experience of objects. (*M&L*, p. 162; emphases added)

This formulation follows a familiar path of avoiding both vulgar materialisms and 'abstracted' idealism in acknowledging these dual dimensions of cultural objects/practices. It also reaches beyond a parallelism of instrumental labour and aesthetic composition. Notation, significantly, here plays a subordinate – albeit crucial – 'instrumental' role to objectification. In this formulation the 'higher' forms of production objectify experiences which include the experience of alienated and unalienated labour ('experience of the production of objects'). Within the cultural materialism, this is an extremely rare reminder of Williams's key conception of the 'communicative' role of art. While it is legitimate for Williams to go on to contest a reductive materialism by insisting that this objectification is a form of 'materialization', the result is a slippage in terminological use between aesthetic 'objectification' and 'materialization'.

These tendencies can be usefully reconceptualized here by employing the subcategories of 'materialization' Márkus develops specifically for the sphere of culture. Rather than Williams's 'abstraction' or 'idealism', Márkus defines *dematerialization* as that process of positing the art object *as* an ideal object such that its material dimensions become regarded as 'the transparent, diaphanous vehicle of significations constituting their essential reality' (Márkus, 1994b, p. 19). Conversely, *rematerialization* in the arts is the process of establishing 'an intentional blockage of relations of signification, in order to self-referentially foreground the signifier, the material medium of communication itself, and for setting free its "energies of semiosis"' (1994b, p. 25).

In short, these two concepts aim to capture the transition that Williams would associate with that from the Romantic to modernist aesthetic projects. Williams's fundamental cultural materialist claim against Arnold – that he fails to reveal the 'material of the process' of culture – is better understood as one made against Arnold's projection from a 'dematerialized' – rather than 'abstract idealist' – conception of culture. Likewise, Williams's *social formalism* is an attempt to mount a theoretical project which learns in part from the modernist aesthetic practice of *rematerialization,* and its comparable legacy within formalist literary scholarship and structuralist theory.[47]

Moreover, it is significant that Williams also deploys the category of notation as a means of setting limits to the Saussurean conception of the sign.[48] Indeed Williams's theoretical 'rematerialization' within his cultural materialism moves towards an historicized semiotics as well as the Marxian formulation above. Where Márkus would hold that aesthetico-cultural objects are principally the products of intellectual labour, the mature Williams sees both cultural objects (or, rather, practices) *and* cultural producers as distinguishable by their internalized 'degree of relative solution of signifying practices' (*SOC*, p. 218).[49] In such theoretical formulations, Williams risks losing the 'inbuilt' normative element of conscious design in Marx's labour-based distinction. However, Williams tends to maintain the norm of 'direct autonomous composition' in his discussion of means of communication as means of production. The gain in Williams's social-formal definition, in his view, is the reduction of the risk of *a priori* separation into abstractly derived 'spheres'.

Finally, it is important to note that Williams draws a fundamental distinction between 'Marx on art and Marx on ideas' (*WICTS*, p. 221). As we shall see, Williams is keen to dissociate aesthetic 'culture' from 'ideology'. However, his considerations of the concept of ideology arise more directly in his reflections on 'Western Marxism' and language and so have been held over for discussion in the next and later chapters.

2.6 Excursus: Williams's 'cultural production' and some apparent 'fellow travellers'

While it is true that Williams did not have access to the more recent Marxian scholarship drawn on in the above exegesis, there seems little doubt that he intended his production paradigm for culture to be understood as one based in Marx's conception of production. As will be argued in Chapter 3, the strongest parallel project lies within the work of Adorno. This Marxian character of Williams's position is a quite crucial means of distinguishing his cultural materialism from 'non-Marxian' uses of the category of 'cultural production' that have developed more or less independent circulation within sociology.

The first of these sociological uses is 'the production of culture perspective', a heuristic characterization of mainly US developments within the sociology of culture understood primarily as a sociology of the arts.[50] This can be dealt with briefly. Cultural production is here understood very descriptively 'in its generic sense to refer to the processes of creation, manufacture, distribution, exhibiting, inculcation, evaluation, and consumption' (Peterson, 1976, p. 10). This remarkably inductive self-characterization of the field resembles the chief failings with which it is charged by its critics – a tendency towards a reductive linearity and positivist empiricism.[51] Clearly this approach has little in common, theoretically, with Williams's cultural materialism.

A second – and more significant – sociological usage of 'cultural production' can be found within the work of Pierre Bourdieu. Bourdieu's project superficially resembles Williams's, and Williams was not only aware of but impressed by Bourdieu's work published during his own lifetime.[52] There is a parallel interest in the social role of the arts informed by a scepticism towards orthodox hierarchies of taste and critical judgement or 'distinction' (Bourdieu, 1984). Bourdieu's critique of the failings of Saussurean 'objectivism' also resembles the critique of Saussure by Vološinov that Williams endorsed.[53] Finally, Bourdieu develops a series of mediating concepts – practice, habitus, field – designed to achieve a broadly similar goal to that of Williams's for his cultural materialism: that is, the reservation of a theoretical space for forms of social action that cannot be adequately accounted for by either subjectivist voluntarism or 'objectivist' determinism. When speaking of the arts, both refer to this range of actions as 'cultural production'.

But the notion of 'cultural production' deployed by Bourdieu is, like the 'production of culture perspective', largely descriptive and not as systematically conceptualized as his more pivotal 'cultural' and 'symbolic' capital.[54] The latter concept was shaped by some of those very tendencies shared with Williams listed above. Bourdieu advocates its usage as a means of overcoming all forms of unwarranted 'distinction' attached to aesthetic objects. 'Economic calculation' is so extended to all goods 'material and symbolic without distinction' (Bourdieu, 1977, pp. 177–8). However, the basis of this economic calculation is deemed by Bourdieu to be the production of scarcity. It is the relative rarity of certain cultural goods (e.g. academic qualifications) that provides their convertibility from cultural into economic capital. As critics have noted, this has little to do with a Marxian conception of capital,[55] although it is consistent with orthodox economic notions of 'marginal utility'.[56]

Now, it is true that Williams has no conceptualization comparable to 'cultural capital' at all. However, this is less an oversight than a difference in theoretical orientation. Williams does have his own principle of convertibility from culture to *economic* capital which is more recognizably Marxian. Rather than Bourdieu's more economically orthodox conception of a cultural market, Williams's key linkage is the Marxian notion of cultural productive forces, especially means of cultural production. The chief 'modern' 'convertibility' Williams envisages is that from a handicraft to industrial capitalist usage of means of communication, especially within what Adorno christened 'the culture industry'. At such points cultural productive forces are either transformed into social productive forces, or enter homologously symmetrical or contradictorily asymmetrical relations with them.

This is not to deny the actual commonalities between Williams and Bourdieu. But their conceptions of cultural production are quite distinct.

Finally, it is important to note another comparative distinction. Williams's work is being increasingly compared also with that of Jürgen Habermas.[57] While it is true that there is a kind of 'communicative ethics' in the early Williams especially, he makes no move towards anything like a Habermasian conception of communicative *action*. In this distinct conception of action Habermas finds his solution to the dilemmas found within the alleged 'expressivism' of the production paradigm's use of the category of objectification (as employed in the exegesis in this chapter). Most especially, labour and production are, for Habermas, inadequate to the task of accounting for *normative* interaction. They are necessarily tied to technicist and instrumental conceptions of action. Hence his need for a separate conception of communicative interaction and for an 'ontological' grounding of this conception in turn in his 'ideal speech situation'.[58]

As we have seen, Williams is quite insistent on the production paradigm's usefulness not only intrinsically but *especially* for the field of 'communicative interaction'. Where Habermas sees the concept of productive force as doomed to a technicism, Williams sees its duplication with the realm of culture as a necessary gain. Moreover, such a conception of means of communication as means of production is a necessary step, in Williams's view, in the advancement of a goal Habermas has more recently reiterated, the development of a democratic public sphere. Unlike Habermas, for Williams the production paradigm is certainly far from 'obsolete'.

3
From Criticism to Critique

The previous chapter focussed on Williams's relationship with 'received' Marxism and its influence on his reception of Marx's work. From at least as early as 1969, however, Williams began comparing his own work with that of some significant 'Western Marxists'.[1] This period of careful reception of parallel projects and redefinition of his own was abruptly interrupted in 1976 by Eagleton's Althusserian critique. Williams's response later the same year met that critique point by point but also recognized the need for a more consolidated announcement of the emerging cultural materialism.[2]

Williams's 1971 acknowledgement of Lucien Goldmann's work, and his recognition of Gramsci two years later in 'Base and Superstructure in Marxist Cultural Theory', are usually seen as announcing the beginning of this rapprochement with the 'Western Marxist' tradition.[3] But an equally strong thematic in both cases is the relationship between literary criticism and sociology. Almost all the conceptual problems and innovations of the 'Base and Superstructure' essay, for example, are discussed within literary analysis. Rather than work with an 'anthropological' conception of culture, Williams there locates the 'true crisis' in cultural theory within the aesthetic sense of culture. Williams seems most concerned in these texts with the methodological implications of abandoning orthodox 'Cambridge' literary criticism including the practical criticism in which he was trained.

Much of the impetus for the English translations of the work of Western Marxists, especially of Louis Althusser, had been undertaken by the journal, *New Left Review* (*NLR*). It was also the publishing site for Williams's own essays on Goldmann and Gramsci, and for Eagleton's critique of Williams and Williams's 'reply'. As mentioned in Chapter 1, in 1968 *NLR*'s editor, Perry Anderson, had published an influential essay, 'Components of the National Culture'. This essay provided a context for Williams's reflections. Its ambitious survey of eight intellectual fields moved from the premiss that there was an 'absent centre' in British intellectual culture: *Britain – alone of major Western societies – never produced a classical sociology* (Anderson, 1968, p. 7). Leavisite literary criticism was the discipline which took up the displaced normative role – and the construct of the social totality – of a critical sociology. Anderson thus accounted for Williams's *The Long Revolution* as 'the most significant work of socialist theory' in Britain of its period, as emerging from within this phase of 'detour' through literary criticism (Anderson, 1968, p. 55).[4]

Anderson's 'absent centre' thesis forms a constant (but not directly acknowledged) background to Williams's 1971 Goldmann lecture. After reiterating Anderson's basic thesis, Williams comments: 'But this is where the central problem between literature and social studies at once arises' (1971, p. 4). Williams then applies the 'absent centre' thesis to Cambridge English's repression of methodological discussion within 'practical criticism'. It is in this context that he welcomes Goldmann's work. In 'Base and Superstructure' the same critical spotlight is turned on 'received Marxist theory'. This is the crucial link between these two 'turning point' essays. Their argument is expanded and developed in *Marxism and Literature* and in the dialogue with *NLR* in *Politics and Letters*.[5]

If one confines oneself to these *NLR* texts, an easy narrative presents itself: Williams carefully considers Goldmann's genetic structuralism but eventually sets it aside for his own somewhat eccentric appropriation of Gramsci's hegemony, so settling accounts with at least vulgar applications of 'base and superstructure' along the way.

However, as we shall see, closer examination of these texts – and of *The Sociology of Culture* – reveals that Williams actually combines elements of Goldmann's project with the Gramscian conception of hegemony. Moreover, complicating matters further is a third, rarely mentioned, Western Marxist option that Williams weighs up with equal seriousness outside the *NLR* publishing locale: the work of the Frankfurt School.[6] Indeed, it is the work of Herbert Marcuse, not Lucien Goldmann, that provided Williams's first opportunity to compare his project with a Western Marxist 'fellow traveller'.

3.1 Entertaining the Frankfurt School: emancipatory critique

> I simply didn't know the Frankfurt School except by incidentals and by account; it was surprisingly late in coming into focus. (*P&L*, p. 260)

Even today, the suggestion of a parallel between Williams's work and that of the Frankfurt School may seem almost absurd to many. It would certainly be so deemed by the cultural studies orthodoxy recounted in Chapter 1. From that perspective, Williams's hostility to 'mass' formulations, whether those practised by Leavis or Adorno, is so well documented that it does not seem possible that he could have had anything in common with the Frankfurt project.[7] Such a position incorrectly assumes that the Frankfurt project is constituted by no more than its work on 'mass culture' and, further, that that work is reducible to a cultural élitism. At the very least, Williams did share with members of the Frankfurt School an interest in a mode of critique of 'high culture' which aimed to maintain its critical dimensions.

The texts which record Williams's initial reception of the translation of some major works of the Frankfurt School support this view. Predictably, Williams does state his 'radical disagreement' with the mass culture thesis but nonetheless confesses to the following: 'A more helpful element of the School's work can be seen in its searching analysis of concepts in *Aspects of Sociology*:[8] some of this is remarkably liberating and challenging and at the very foundations of the subject' (Williams, 1974b). It is hardly surprising that Williams found this 'searching analysis of concepts' so helpful. Adorno and Horkheimer (in *Aspects*), Marcuse (and, later, Habermas) all practise what often seems an identical mode of historical semantics to Williams's. Marcuse, for example, published an essay in 1965 called 'Remarks on a Redefinition of Culture' which bears an uncanny resemblance to Williams's historical semantic discussions of the concept.[9]

Williams appears to have been unaware of that essay but asserted very strong affiliations with some of Marcuse's work in the longest of his reviews of Frankfurt School publications. This was a 1969 review essay on Marcuse's *Negations*, a collection of republished articles from the 1930s to 1960s. Despite his own comparable disappointments, Williams distances himself from Marcuse's position on 'the condition of the proletariat in advanced capitalist society' by, significantly, questioning the degree of Marcuse's association with 'an American sociology which, even in its most critical and even revolutionary forms, appears to me distorted by the very pressures and contradictions of its society'. Yet Williams goes on to indicate more fundamental common interests in this remarkable passage:

> My interest, and it is deep, is in what might be loosely called the German rather than the American work, and this is in fact predominant in *Negations*. For historical reasons, we have been separated, in Britain, from a critical and philosophical tradition which, when we re-encounter it in Marcuse or in Lukács, is at once strange and fascinating: at once broader and more confident, more abstract and yet more profoundly involved than our own. I felt the size of this gap, and yet the interest and pleasure of a possible bridge across it, in one of Marcuse's essays from the thirties... on 'The Affirmative Character of Culture'. The particular interest of the essay, for me, is that its analysis corresponded so closely with a central theme of *Culture and Society*, and that both were historical treatments, of very much the same problem, which were yet continents of countries apart in method and in language. It was a marvellous moment of intellectual liberation to read across that gap into a mind which in all but its most central area of concern and value was so wholly other and strange.

> 'Affirmative culture was the historical form in which were preserved those human wants which surpassed the material reproduction of existence'.[10]
>
> This was exactly my own conclusion, of the essential origin and operation of the idea of culture, as it developed in England after the Industrial Revolution, at a time when we were very close, especially through Coleridge and Carlyle, to the German thought to which Marcuse's arguments relate. It is a sense of meeting after a long separation.... it indicates in a very sharp and uncompromising way an issue that has been at the centre of my own concern since I returned to Cambridge: the social and political use of what appears to be the ideal or the beautiful content of what Marcuse calls 'affirmative culture'. That is, an idea of culture represented human values which the society repressed or could not realize. As such it was critical. But the form of the separation became at a certain point (in England, perhaps, in the late nineteenth century, when the ethos of what we call traditional Cambridge was formed) a ratification, a system of values against social involvement and social change. (Williams, 1969a, p. 368)

Williams identifies 'the use made of the reconciling group in practical criticism and the more openly ideological use of a late nineteenth century idea of tragedy' as British examples of such an affirmative usage of culture. The sentence Williams cites from Marcuse strongly echoes Williams's conclusion concerning 'unrealized possibilities' in 'The Analysis of Culture'. The whole passage thus does far more than acknowledge a commonality between Marcuse's 1937 essay and *Culture and Society*. Despite its aside concerning *initially* different methods, it provides a different route 'out' of Cambridge literary criticism towards a similar (to Marcuse's) 'central area of value and concern'. It constitutes a near recognition by Williams that he shares with members of the Frankfurt School a key practice of their Critical Theory: 'emancipatory' ideology critique.

Perhaps because it bridges conceptions of 'critique' and 'ideology', this 'method' remains relatively unknown in much English language commentary outside specialist literature on the Frankfurt School and Critical Theory. Its relevance to recent related debates is considerable for, as Habermas has recently argued, ideologies deemed susceptible to such critique 'differ from Foucaultian discourses because of their capacity for self-transformation' (Habermas, 1993, p. 429). Some elaboration here is thus required.[11]

Broadly, by such ideology critique is meant an immanent critique of an ideology according to its own inner standards where 'ideology' refers primarily to 'elaborated' ideologies (broadly, philosophies and theories) but the technique can be applied also to aesthetic works. The often utopian claims of such ideologies, their emancipatory promise, are seen to derive from their 'content' but are subject to socio-historical delimitation which is traceable to 'external' determinants but also immanently present in 'closures of thought' which contradict the ideology's broader emancipatory claims.[12]

Crucially, this conception usually restricts the very usage of the concept of ideology to, in effect, works of 'high culture' (including elaborated political ideologies).[13] Adorno draws this distinction especially clearly in *Aspects of Sociology*:[14]

> Accordingly, the critique of ideology, as the confrontation of ideology with its own truth, is only possible insofar as the ideology contains a rational element with which the critique can deal. That applies to ideas such as those of liberalism, individualism, the identity of spirit and reality. But whoever would want to criticize, for instance, the so-called ideology of National Socialism would find himself victim of an impotent naiveté. Not only is the intellectual level of the authors Hitler and Rosenberg beneath all criticism. The lack of any such level, the triumph over which must be counted among the most modest of pleasures, is the symptom of a state, to which the concept of ideology, of a necessarily false consciousness, is no longer directly relevant . . . rather it is a manipulative contrivance, a mere instrument of power, which actually no-one, not even those who used it themselves, ever believed or expected to be taken seriously. With a sly wink they point to their power: try using your reason against that, and you will see where you end up; . . . Where ideologies are replaced by approved views decreed from above, the critique of ideology must be replaced by *cui bono* – in whose interest? (Frankfurt Institute For Social Research, 1973, p. 190)

The 'truth' of genuine ideologies lies in their promise; their 'falseness' in the legitimating pretence that such promise has *already* been fulfilled.[15] Crucially, however: 'For ideology in the proper sense relationships of power are required which are not comprehensible to this power itself, which are mediated and therefore also less harsh' (Frankfurt Institute for Social Research, 1973, p. 191). Nazism fails this test, on Adorno's account, because it has no legitimating ideology comparable to the utopian promise of bourgeois ideals of justice and democracy. In this sense, ideologies are also embodiments of a 'false consciousness' *amongst ideologists and those with power* (but not necessarily those so dominated) yet this falseness, like their truth, is 'necessary' because of these mediated power relations. Nazism lacked such mediations. The exposure of the social interests behind such consciously 'unmediated' ideologies can be distinguished from emancipatory ideology critique – following Márkus – as an 'unmasking' critique.[16]

Adorno, in contrast, simply reserves the concept of ideology for its emancipatory sense in the passage above. For the criticism of individual works of art, he provides a similar account which is usually described as relying on two stages, immanent and transcendent critique.[17] Art and philosophy share a delegitimating 'truth content' that immanent analysis can draw out in contrast to the 'necessary false consciousness' of ideology.[18] Here Adorno discusses this conception of critique in relation to lyric poetry:

> the social interpretation of lyric poetry as of all great works of art ... must discover how the entirety of a society, conceived as an internally contradictory unity, is manifested in the work of art, in what way the work of art remains subject to society and in what way it transcends it. In philosophical terms, the approach must be an immanent one. Social concepts should not be applied to the works from without but rather drawn from an exacting examination of the works themselves. Goethe's statement in his *Maxims and Reflections* that what you do not understand you do not possess holds not only for the aesthetic attitude to works of art but for aesthetic theory as well; nothing that is not in the works, not part of their own form, can legitimate a determination of what their substance, that which has entered into their poetry, represents in social terms. To determine that, of course, requires both knowledge of the interior of the works of art and knowledge of the society outside.... The greatness of works of art ... consists solely in the fact that they give voice to what ideology hides. Their very success moves beyond false consciousness, whether intentionally or not. (Adorno, 1991a, pp. 38–9)

Adorno's approach to aesthetic works thus shares much with Williams's practice of 'looking both ways' discussed in the introductory preface to this book. It is in analytic pursuit of the contradictory 'pretensions' of the embedded truth content within the work of art that immanent critique 'perceives those [antinomies] of society' (Adorno, 1984a, p. 32). This consequence of immanent cultural critique in turn necessitates a complementary move to an 'external' transcendent critique, including possibly the terrain of base and superstructure. However, the radical historicization demonstrated by Adorno in the earlier passage above is equally demanded at this stage – that is, the form of transcendent critique chosen must be appropriate to the historical conjuncture. Adorno was famously pessimistic about the role of rationalizing tendencies of capitalism as manifested in 'the culture industry'. These historically delimited the capacity for autonomy not only in commodified 'mass culture' but equally in 'high culture'. In short, art too was in danger of becoming unworthy of critique, another 'manipulative contrivance'.

We saw in Section 1.4 that Williams developed an undeclared mode of immanent critique as a subversion of his practical critical training in literary criticism. The critique of Arnold reconstructed there fulfils Marcuse's conception of affirmative culture, as Williams himself recognized, but *also* fulfils Adorno's conception of emancipatory critique.

Perhaps the most relevant parallel case to Williams's methodological transition is none other than that of Marx. The establishment of the distinction between 'mere criticism' and 'critique' was a key task for him during 1843–44. The former for Marx – exemplified by the Young Hegelians – refers to the application of arbitrary *external* standards to the object of criticism, so risking a decline into *a prioristic* dogmatism. Critique, in contrast, recognizes a contradictory tension of actual and possible such that the 'object' is not considered merely an inert object at all. As Benhabib puts it:

> Marxian critique... is not a mode of *criteriological* inquiry. The criteria it presupposes in its inquiry are not different from the ones by which the object or phenomenon judges itself. The Marxian method of critique presupposes that its object of inquiry is reflexive; it presupposes that what is investigated is already a social reality which has its own self-interpretation. (Benhabib, 1986, p. 33)

Williams too rejected 'mere criticism' for critique but this criticism was literary rather than philosophical. The literary criticism that he rejected had applied ostensibly externally fixed (but slippery) standards such as Arnold's 'best'. As we saw in Chapter 1, Williams effectively restored the 'ideal' Enlightenment dimension to the 'British' conception of culture. It is the acknowledgement of the ossification of the concept of culture into the 'standards' of 'criticism' that Williams recognized as the common ground of *Culture and Society* and Marcuse's 1937 essay.

However, Williams never again openly embraced or even acknowledged Marcuse's or Adorno's conception of critique as such after his essay on Marcuse.[19]

In a very recent debate that brings this Frankfurt School practice directly into dialogue with current cultural studies orthodoxy, Francis Mulhern has recently reasserted a central dilemma faced by both Adorno and Marcuse.[20] Their practice of immanent critique necessarily took place in an absence of any likely 'transcending' complement. Immanent critique might thus produce its immanent 'truth' but, in the absence of any likely agent of transcendent social change, Adorno especially was so 'not spared the general curse of regression' into 'the natural aristocratism of Kulturkritik', Mulhern's term for the conservative cultural criticism of Germany and Britain that would include those I have categorized as 'clerisists' (Mulhern, 2002, p. 96).

As argued in Chapter 1, Williams faced his own loss of an 'enabling social subject' in the 1960s yet not only maintained his practice of critique but avoided Adorno's 'curse'. Williams appears to have drawn a similar conclusion to Mulhern's about Adorno and Marcuse's *later* work – by which he appears to have meant their late works on aesthetics – and frequently confines his expressions of respect for the work of the Frankfurt School to that of the 1930s. As we shall see in Section 3.3, he also drew on Adorno's and Benjamin's discussions of modes of correspondence in developing his 'social formalism' and came close to recognizing that he shared their interest in the production paradigm. Crucially, it was by this means – the recognition of the complexity of modes of correspondence of cultural productive forces and relations of cultural production – that Williams escaped the 'curse of regression' into Mulhern's Kulturkritik. That is, Williams – and even to a certain degree Adorno – demonstrated a preparedness to recognize varying degrees of autonomy in popular cultural practices that open the prospect that emancipatory critique might be extended beyond any apparently élitist confinement to 'high culture'.[21]

Williams so 'moved on' to a more historicized and nuanced conception of emancipatory critique. The first step towards this was his more detailed discussion of Lucien Goldmann's sociology of literature.

3.2 From Goldmann to Gramsci?

The work of Lucien Goldmann is undoubtedly a chief point of reference for Williams's adoption of what he called the 'modern Marxist conception of homology' in 'Marx on Culture' (Section 2.3). Homology emerges as the key term in Goldmann's later reformulations and advocacy of his *genetic structuralist* method. The homology Goldmann proposes as the core of this method is posited to exist between 'world views', authors and their literary works. Here is one of his own most succinct accounts:

> its basic hypothesis being precisely that the collective character of literary creation derives from the fact that the *structures* of the world of the work are homologous with the mental structures of certain social groups or is in intelligible relation with them, whereas on the level of content, that is to say, of the creation of the imaginary worlds governed by these structures, the writer has total freedom. The use of the immediate aspect of his individual experience in order to create these imaginary worlds is no doubt frequent and possible but in no way essential and its elucidation constitutes only a useful, secondary task of literary analysis.
>
> In reality, the relation between the creative group and the work generally appears according to the following model: the group constitutes a process of structuration that elaborates in the consciousness of its members affective, intellectual and practical tendencies towards a coherent response to the problems presented by their relations with nature and their interhuman relations. With few exceptions these tendencies fall far short of effective coherence…
>
> Furthermore, mental categories exist in the group only in the form of tendencies moving towards a coherence I have called a world-view, a view that the group does not therefore create, but whose constituent elements it elaborates (and it alone can elaborate) and the energy that

> makes it possible to bring them together. The great writer (or artist) is precisely the exceptional individual who succeeds in creating in a given domain, that of the literary (or pictorial, conceptual, musical etc.) work, an imaginary, coherent, or almost strictly coherent world, whose structure corresponds to that towards which the whole of the group is tending; as for the work, it is, in relation to other works, more or less important as its structure moves away from or close to rigorous coherence. (Goldmann, 1986, pp. 159–60)

However, in his early work Goldmann does not employ the category of homology to describe these structural correspondences but instead uses a revised understanding of the Lukácsian conception of 'totality' and refers generally to a 'dialectical method'. Goldmann eventually transformed his dominantly Lukácsian vocabulary into that of Piaget's genetic epistemology and 'structuralism'. Piaget's conception of structuralism also allowed Goldmann to sustain a Lukácsian 'holism' against the growing influence of Saussurean structuralism. In particular, Piaget challenged the delimited 'analytic structuralism' of Lévi-Strauss which conceived of structure as aggregates of component elements such as 'mythemes'. Piaget advocated instead the necessity of a wholistic perspective.[22]

Indeed, the concepts of homology and formal correspondence are known more widely for their role within 'the linguistic model' of formalist-structuralist analysis. Principally derived from the leading Prague formalist, Roman Jakobson, and applied to kinship systems and myths by Lévi-Strauss, they are usually taken to refer to the establishment of formal resemblances between two sets of binary oppositions (or 'differences'). For example, in Lévi-Strauss's interpretation of the resemblance by association made by The Nuer people between twins and birds, he says: 'It is not the resemblances but the differences which resemble each other' (Lévi-Strauss, 1973, p. 149).[23] That is, the resemblance is not to be found superficially present in the semantic 'content' of 'twins' or 'birds' but formally, in the system of differentiation within which 'twins' and 'birds' are positioned:

> Twins "are birds", not because they are confused with them or because they look like them, but because twins, in relation to other men, are as "persons of the above" are to "persons of the below", as "birds of the below" are to "birds of the above". (Lévi-Strauss, 1973, p. 153)

The linguistic model of formal correspondence so claims to render comprehensible associations otherwise unintelligible to the outsider.

Neither Goldmann nor Williams, however, ever worked with this binary-based understanding of homological correspondence. Indeed, because of the centrality of Lévi-Strauss's work to the successful influence of structuralism outside linguistics and anthropology – and as a point of departure for poststructuralism[24] – it is a matter of some controversy as to what Goldmann's homologies, in contrast, claim to achieve. Like Piaget, he holds to a different conception of 'structure'. Correspondence for Goldmann takes place between the structuration of the world view of the group and the 'mental categories' of the 'coherent' literary work.

Accordingly he is prone to the charge that his homologies are correspondences of formless 'content' and 'class structure', so merely producing a transposed vulgar application of the base and superstructure metaphor.[25] A more accurate line of criticism, of which Williams's was one of the first examples, stresses the tension between Goldmann's early and later work.

Williams argued that Goldmann was successful in proving his thesis, in the extended 'epochal' case in what is generally agreed to be his masterwork, *The Hidden God*. Goldmann's key 'structure' in that work is his 'signifying structure' that remarkably resembles Williams's 'structure of feeling'. Goldmann argues that Pascal's *Pensées* and Racine's tragedies share the signifying structure of a tragic innerworldly refusal manifested by decisions taken by their acting subjects to 'consciously reject a world in which they are compelled to live' (Zima, 1999, p. 91).[26] However, he does not 'reduce' this signifying structure to an 'interest' of his 'social group', *the noblesse de robe*. Rather, he finds in Jansenism's theological discourse a mediating 'world view' that functions similarly for that group. It legitimated the *noblesse de robe*'s withdrawal from public life necessitated by their political abandonment by Louis XIV. Thus Jansenism performed a function for the *noblesse de robe* that was 'enacted' in Racine's tragedies. It is in this sense that Goldmann practises 'functional homologies'.

Critics other than Williams also stress how Goldmann's later analyses, notably the more micro-analytic projects such as his analysis of Malraux's novels, fail. This failure is partly because Goldmann cannot convincingly establish an appropriate social group's world view from which to generate a corresponding functional homology and also because his presumption of coherence cannot cope with avant-gardist practices.[27]

This criticism is best formulated theoretically by Fredric Jameson. Goldmann's later homologies constitute for Jameson a mechanistic failure in an attempted development of an adequate methodologization of *mediation*, 'the classical dialectical term for the establishment of relationships between, say, the formal analysis of a work of art and the social ground…' (Jameson, 1981, p. 39).

Although he later expressed significant differences with Adorno, Goldmann's own 'dialectical method' of analysis of literary works shares much with Adorno's conception of immanent emancipatory critique, most obviously in its view that only 'coherent' examples were worthy of a full analysis.[28]

Likewise Goldmann's conception of structuration of world views strongly resembles that which Jameson attributes to Marx's '*Brumaire* solution', a 'structural limitation'. Neither Goldmann, Jameson nor Williams, however, draw this connection between Goldmann and *The Brumaire* explicitly.

Perhaps Williams and Goldmann's strongest common ground is a conception of homology as the reproduction of 'limit setting' as 'closures in thought'. Yet it is also possible to see a remarkable convergence between the very 'content' of Goldmann's signifying structure of 'innerworldly refusal' and Williams's structure of feeling of 'liberal tragedy' in *Modern Tragedy*. Each appears to have had a specific interest in developments within the tragic form that help to account for the self-incapacitation of the public life of a whole social group.[29]

Williams also regards Goldmann's work as an interesting failure in mediation but nonetheless one worthy of reconstruction from Goldmann's own primary categories. The latter included a conception of a 'collective subject' and a related distinction between actual and potential (or 'possible') consciousness based on Lukács's distinction between empirical and imputed consciousness and, also, between false and true consciousness.[30] Significantly, Williams does not challenge the necessary implication here of 'false consciousness':

> Goldmann, following Lukács, distinguishes between actual consciousness and possible consciousness: the actual, with its rich but incoherent multiplicity; the possible, with its degree of maximum adequacy and coherence. A social group is ordinarily limited to its actual consciousness, and this will include many kinds of misunderstanding and illusion: elements of false consciousness which will often, of course, be used and reflected in ordinary literature. (1971, p. 11; *PMC*, p. 23)[31]

Here the conception of determinant 'limit' meets Williams's sense of determination as 'pressure'. The 'limit' of 'actual consciousness' is susceptible to a critique that reveals 'potential consciousness'. The dynamic of a Frankfurt ideology critique between 'actual' and possible' is so (re)established. The more immediate appeal of this model for Williams was its capacity to break Cambridge English's neat distinction between a 'background' world view (e.g. the Elizabethan) and works of literature. Williams's own work suggested a 'baffling' non-correspondence between literary works and these 'backgrounds'.

However, Williams expressed doubts about Goldmann's exclusive focus on the analysis of the coherent work as evidence of 'potential consciousness', while restricting actual consciousness to a somewhat doctrinal conception of world view. Williams's own work, focussed on moments of social transition, had, he believed, discovered something else:

> There were real social and natural relationships, and there were relatively organized, relatively coherent formations of these relationships, in contemporary institutions and beliefs. But what seemed to happen, in the greatest literature,[32] was a simultaneous realization of and response to these underlying and formative structures. Indeed, that constituted, for

> me, the specific literary phenomenon: the dramatization of a process, the making of a fiction, in which the constituting elements, of real social life and beliefs, were simultaneously actualized and in an important way differently experienced, the difference residing in the imaginative act, the imaginative method, the specific and genuinely unprecedented imaginative organization. (1971, p. 12; *PMC*, p. 24)

Indeed, Williams is too modest here. His own dialectical method in *The Long Revolution* had already overcome many of Goldmann's failings. His analysis of the reasons for the unevenness of Goldmann's work quietly draws on the lessons of 'The Analysis of Culture'. He attributes this unevenness to two key elements in Goldmann's method. First, the class-based limits of 'possible consciousness' constrict Goldmann's capacity to deal with smaller-scale social transformations than those of the class dominance of an 'epoch'. Second, there is a related tendency to conceptualize cultural forms as permanent trans-historical phenomena. Williams referred to these problems generically as the privileging of epochal over historical analysis. As we saw in Chapter 1, 'The Analysis of Culture' aimed to overcome such failings.

As in 'The Analysis of Culture', Williams poses 'the problem' as one of method:

> The problem is always one of method, and this is where . . . [Goldmann's] idea, of the structures of the genesis of consciousness, must be taken very seriously.[33] We are weakest, in social studies, in just this area: in what is called the sociology of knowledge but is always more than that, for it is not only knowledge we are concerned with but all the active processes of learning, imagination, creation, performance. (1971, p. 15; *PMC*, p. 29)

This comment appears consistent with the criticism – made most forcefully by Zima – that Goldmann's focus on quasi-philosophical 'coherence' prevents any recognition of the polysemeity of literary works; that is, their capacity to enable different modes of interpretation in different social and historical contexts.[34] This problem, for Williams, was resolvable by a radical historicization of the concept of cultural form.[35] Here, as he states explicitly in the previous citation, Williams wishes to highlight his own understanding of the 'specifically literary phenomenon' understood as 'the dramatization of a process'.

In a second assessment of Goldmann a year later, Williams developed his methodological interest one stage further by explicitly linking Goldmann's emphasis on consciousness with '15 years' of his own work on the relationship between cultural forms and (this time eschewing 'structures of feeling') the emergent creative consciousness of a 'generation' of creative practitioners (Williams, 1972, p. 376). It is this emphasis on the emergent that provides one of the strongest 'positive' links between Williams's reception of Goldmann and that of Gramsci in 'Base and Superstructure' in 1973.

However, the initial connections made between Goldmann and Gramsci within that later article are certainly strictly corrective of the former. 'Totality' is explicitly set aside for hegemony, because of Williams's desire to maintain a delimited ideological superstructure with a class character. In an even more explicit allusion to Goldmann, Williams argues that hegemony offers a corrective to the tendency for the 'best Marxist cultural analysis' to privilege epochal over historical analysis. It is at this point that Williams introduces his famous distinction between dominant, residual and emergent practices and forms within a contingent hegemony (summarized in Table 3.1). In this context, it can be seen also as an initial corrective to Goldmann's 'permanent' cultural forms.[36]

Table 3.1 Key features of Williams's account of hegemony[a]

Position of socio-cultural practice/form	**Definition/role in Hegemony**	**Example**
Dominant	Central system of meanings and values which is dependent for renewal on process of incorporation of elements of residual and emergent forms. Agencies of incorporation are primarily 'socializing' institutions, selective traditions and formations (informal artistic/ intellectual groupings)	British hegemony in a given period
Residual	Formerly dominant forms which have survived to play a reduced but active role at present (unlike the fully incorporated archaic). May assume incorporated, alternative or oppositional role towards the dominant	Idea of rural community; organized (Christian) religion
Emergent	New forms whose most likely sources are a rising class, new formations or new social movements. May assume incorporated, alternative or oppositional role towards the dominant	Nineteenth-century British radical popular press (which moved from oppositional to incorporated)[b]
Pre-emergent/ structure of feeling	Pre-articulated 'social experiences in solution' at a stage prior to their achieving an objectivated form	That which is (later) rendered in historical semantic shifts in '*Keywords*'

[a] This table draws on materials by Williams beyond the argument of the 'Base and Superstructure' essay.
[b] *M&L*, p. 124; cf. Williams (1978b).

Williams certainly accepts Gramsci's primary distinction between non-hegemonic rule entirely by coercion and hegemonic rule by consent backed by coercion.[37] However, he rejects any suggestion that such consent is merely subordination to a fully formed ideology or even to Goldmann's 'world-view'.[38] As he later puts it, 'hegemony is not only the articulate upper-level of ideology' (*M&L*, p. 110). Nor is some form of manipulation or even persuasion of popular consciousness the key mechanism of hegemonic rule for Williams. Yet, precisely because it is more than this, a successful hegemony can be described as 'deeply saturating the consciousness of a society' (*PMC*, p. 37), so that the 'pressures and limits' of a particular social order 'seem to most of us the pressures and limits of simple experience and common sense' (*M&L*, p. 110).

The key hegemonic mechanism for Williams is the *incorporation* of practices and forms that emerge outside the control of the dominant and on which the dominant is dependent for renewal. In his eagerness to contest contemporary interpretations of Gramsci (most notably Althusser's), there is a tendency for Williams to re-run his critique of 1930s English Marxism and of Eliot's and Hoggart's class reductivisms. He even re-employs phrases like 'co-operative shaping and common contribution' in configuring his alternative (*M&L*, p. 112). However, unlike his arguments of the late 1950s and early 1960s, there is no direct invocation of democratic working-class institutions as a given bearer of what is here called an alternative hegemony. Instead, he quite explicitly moves *beyond* his own former (and Gramsci's) emphasis on 'the coming to consciousness of a new class' as the sole source of emergent cultural practice (1973a, p. 12; *PMC*, p. 42):[39]

> We have then one central source of new practice, in the emergence of a new class. But we have also to recognize certain other kinds of source, and in cultural practice some of these are very important. I would say that we can recognize them on the basis of this proposition: that no mode of production, and therefore no dominant society or order of society, and therefore no dominant culture, exhausts human practice, human energy, human intention. (1973a, p. 12)[40]

These alternatives can include 'alternative perceptions of others, in immediate personal relationships, or new perceptions of material and media, in art and science' (*PMC*, p. 44). This listing could be described as characteristically open or vague but in hindsight it does seem to anticipate Williams's later explicit recognition of feminism and other social movements, the role of aesthetic avant-gardes and other 'formations' and the significance of the social design of new means of communication.

Given Williams's deliberate move beyond the warrant of a central part of Gramsci's position, just how consistent his employment of hegemony is with Gramsci's own work is a moot point. Williams is certainly justified in reclaiming the distinction between ideology and hegemony from Althusser's conflation of the two.[41] A case can also be made that Williams's emphasis on incorporation is consistent with Gramsci. Gramsci's account of the relationship between organic and traditional intellectuals, for example, suggests such a model, as does his reminder that all hegemonic orders are contingently dependent on 'unstable equilibria' of social forces.[42]

Gramsci's admiration for Marx's *Brumaire* was noted in Chapter 2. But what Gramsci also adds to *The Brumaire* analysis is a developed account of the *social production* of ideologies, 'organically' within social classes and/or 'blocs' of social forces, and their 'elaboration' in civil society. 'The Formation of the Intellectuals' is remarkably consistent with '*The Brumaire* solution'.[43] The 'internal' social production of organic ideologies is a prerequisite for any successful hegemonic rule – a claimed 'general interest' – by an 'historic bloc' formed chiefly from an alliance of class fractions.[44]

While Williams does not acknowledge this aspect of Gramsci directly, his case study, 'The Bloomsbury Fraction', explicitly resembles it. As mentioned in Chapter 2, this analysis perfectly echoes Marx's in *The Brumaire* of the petty bourgeois representative. Yet it would appear also to address absences in Goldmann's genetic structuralism which was committed to establishing, as Williams put it, 'the historical... formation and development of *structures* (forms of consciousness)' (*KW2*, p. 143). In this sense both Goldmann's and Gramsci's methods were 'genetic'.

The chief difference between 'The Bloomsbury Fraction' and *The Brumaire*, however, is that Williams *reverses* the famous determinant role granted in *The Brumaire* to signifying traditions: the 'conjuring up of the spirits of the past' and their costumes as the preferred mode of representation of new social forces (Marx, 1958b, p. 225). Rather, Williams proposes the following thesis concerning 'bourgeois fractions' during a comparative discussion of the pre-Raphaelites and the Bloomsbury group in the same essay:

> But in their effective moment, for all their difficulties, they were not only a break from their class – the irreverent and rebellious young – but a means towards the necessary next stage of development of that class itself. Indeed this happens again and again with bourgeois fractions: that a group detaches itself... in terms which really do belong to a phase of that class itself, but a phase now overlaid by the blockages of later development. It is a revolt against the class but for the class, and it is really no surprise that its emphases of style, suitably mediated, become the popular bourgeois art of the next historical period. (1978a, p. 54; cf. *PMC*, p. 159)

The prefigurative character of 'bourgeois dissidence' becomes a key device in Williams's sociological account of the limitations of the many 'modern' intellectual formations he challenged. Their apparent innovation is often revealed to have a conservative fate of re-incorporation into the hegemonic ideology of the (ruling) class from which they broke. The primary sociological task thus becomes that of establishing the conditions which contribute to the particular dissident ethic within the intellectual formation and thus the further provision of its 'limits', the key to its homologous relation. Their autonomy, while real and for Williams open to detailed sociological description, is almost necessarily temporary. The elements of this newly formed 'ethic' tend to move from the emergent to alternative to incorporation within the dominant as a component of a revised legitimating ideology.[45]

The resemblance with Goldmann's genetic structuralism is thus also striking. The adoption of a class fractional analysis answers Williams's major complaint concerning Goldmann's tendency to epochal analysis.

Likewise, this model of analysis of intellectual formations is quite consistent with Gramsci's emphasis on the social production of 'organic ideologies' within hegemonic ruling blocs. However, where Gramsci sees such ideologies as the social cement that binds the membership at the initial formation of such a bloc, Williams's analysis here assumes the pre-existence of an enduring hegemony. His class fractional analysis seeks to explain the origins of a dissidence from members within a ruling bloc that survives independently for a time but is vulnerable to incorporation.

Thus, rather than move from Goldmann to Gramsci, Williams repositioned Goldmann's focus on 'structures of the genesis of consciousness' within an adapted version of Gramsci's hegemony.

However, while 'The Bloomsbury Fraction' gives some indication of Williams's greater historicization of Goldmann's genetic method, it still leaves the question of cultural forms in abeyance. This need required Williams to clarify further his conception of mediating correspondences. In spite of his later homage to the Russian *social* formalists (Chapter 4), he found appropriate theoretical support on this occasion in the work of Adorno and Benjamin.

3.3 Adorno and Benjamin: mediation, cultural productive forces, correspondence

As with the earlier discussion of the *NLR* 'turning point' essays, a superficial reading of the chapters immediately preceding 'Hegemony' in *Marxism and Literature* would suggest that Williams's comparative assessment of key Western Marxist figures resulted in the setting aside of all those he discusses but Gramsci. However, the movement of his argument is actually far more nuanced.

With the orthodox reflectionist models rejected once again, Williams turns to the possibility of a 'mediating' relationship between base and superstructure where 'mediation' is initially defined as 'an *indirect* connection or agency between different kinds of act'. Again, he quickly sets aside as unsatisfactory a negative version of mediation as 'indirect expression' in which an ideological process of 'masking' takes place. Analysis of such mediation would thus be 'a process of working back through the mediation to their original forms' (*M&L*, p. 98).[46] Williams rejects such conceptions of mediation as they rely on an *a prioristic* dualism analogous to a reductivist deployment of base and superstructure.

He expresses far more interest in what he calls 'the contribution of the Frankfurt School' to a 'positive' understanding of mediation. However, he only refers directly to Adorno's 'Theses on the Sociology of Art':[47]

> Here the change involved in "mediation" is not necessarily seen as distortion or disguise. Rather, all active relations between different kinds of being and consciousness are inevitably mediated, and this process is not a separable agency – a "medium" – but intrinsic to the properties of the related kinds. "Mediation is in the object itself, not something between the object and that to which it is brought."[48] Thus mediation is a positive process in reality, rather than a process added to by way of projection, disguise or interpretation. (*M&L*, pp. 98–9)

Williams does not develop his interest in 'positive mediation' any further at this point. The phrase he approvingly cites from Adorno's 'Theses on the Sociology of Art' – 'mediation in the object itself' – is part of a contrast Adorno makes in a reply to a critic of his *Introduction to the Sociology of Music*.[49] He contrasts a positivist effects-based 'communication' model with an emancipatory critique. He continues thus shortly after the passage cited by Williams:

> What I mean, in other words, is the very specific question aimed at products of the mind, as to how social structural moments, positions, ideologies and whatever else, assert themselves in the work of art themselves. I brought out the extraordinary difficulty of the question quite deliberately and without reducing it, and thereby the difficulty of a sociology of music which is not satisfied with external arrangements, not satisfied with the position of art in society, with the effects it has in society, but wants to know how society objectivates itself in works of art. (Adorno, 1972, p. 128)

Mediation refers here, in effect, to the preconditions of an emancipatory critique. As we saw earlier above, in his discussion of ideology in *Aspects of Sociology*, Adorno employs the category of mediation to capture the very distinction between an ideology worthy of such critique and the 'transparent' use of power by Nazism. Williams thus effectively acknowledges Adorno's conception of emancipatory critique as his 'positive conception of mediation'.

The discussion of mediation in *Introduction to the Sociology of Music* to which Adorno refers in his 'Theses on the Sociology of Art' is regarded as perhaps his most pivotal by scholars of Adorno.[50] Adorno argues there for an alternative to dualism in any usage of mediation which Williams would have recognized from his advocacy of a production paradigm: the conceptualization of artistic objectivation as related to a form of 'basal' *production* grounded in a common 'social labour'.[51]

The category of cultural productive force – a culturally specified adaptation of part of the traditional Marxian 'base' – is thus fundamental to both Williams's and Adorno's projects as a mediating category. It played a key analytic role for Adorno from his earliest (1932) article for the Frankfurt Institute, 'On the Social Situation of Music', to his posthumously published *Aesthetic Theory*.[52] There are two prime fields of application of the category of cultural productive force for Adorno: the 'progressive' deployment of specifically aesthetic powers by the creative subject and the use of existent technics in aesthetic production by that subject. To these can be added the knowledge of genres (aesthetic cultural forms) and other necessary intellectual 'skills' as constitutive of the artist's creative practice within aesthetic 'autarky'.

Adorno sees a risk, though, in the usage of certain of these productive forces in that the necessary objectivations of the aesthetic product may become mechanically fetishized[53] – or the progress of aesthetic productive forces can become dependent on the alienated social productive forces with which they become entwined; that is, technologies required in performance and recording. This position provided the basis for Adorno's famous disagreements with Walter Benjamin, especially those provoked by the draft of Benjamin's essay, 'The Work of Art in the Age of Mechanical Reproduction'. Adorno countered Benjamin's relative optimism about the 'progressive' consequences of photography and cinema's dependence on new 'mechanical' means of cultural production.[54] One issue at stake in this debate is the very distinction between aesthetic 'technique' and the 'technology' of reproduction in the then 'new technologies' and, especially in the case of film, whether these were separable and, if so, whether they were qualitatively different.[55] Benjamin's optimism rested, in Adorno's view, in a confusion of the two and a related failure to recognize the role of the culture industry. Williams understood this distinction similarly as one between the uses of means of communication as means of cultural production and as means of general production. Unlike Adorno, however, he saw a greater range of determinate possibilities within these.[56]

It is ironic that Williams came so close to recognizing the commonality between his own use of the production paradigm and that in Adorno's discussion of his *Introduction to the Sociology of Music*, but nonetheless missed it. However, his near silence on Benjamin's far better known 'Work of Art' essay – which directly addressed so many of his interests – is very curious.[57]

In contrast, Williams showed explicit interest in the then available materials from Benjamin's draft for his study of Baudelaire – which in turn formed part of his uncompleted *Arcades Project (Passagen-Werk)* – and the related correspondence between Benjamin and Adorno. This correspondence is now also famous for Adorno's less than wholly sympathetic assessment of Benjamin's material.[58]

The two drafts on which Adorno commented, 'Paris, Capital of the Nineteenth Century' (the 1935 '*exposé*') and 'The Paris of the Second Empire in Baudelaire' (1938), and the text finally published by the Frankfurt School's Institute for Social Research in 1939, 'Some Motifs in Baudelaire', were published in English together in 1973. Williams had reviewed this book the same year.[59] He there made plain his lack of interest in the last of these Benjamin texts, not so much because of the effects of Adorno's seemingly 'disastrous' critiques, but because the result resembled a tendency towards 'a sophisticated late form of idealism' common within contemporary 'cultural studies' (Williams, 1973b, p. 22).

While this comment is somewhat enigmatic, it is not difficult to see why Williams preferred the first two 'stages' and characterized them as 'indispensable as well as brilliant'. For Benjamin demonstrated, especially in his second stage – his expansion of 'Baudelaire and the Streets' from the first 1935 *exposé* – a method that much resembled one Williams had developed himself, especially in his studies of Dickens within *The English Novel: from Dickens to Lawrence* and *The Country and the City*.[60] Williams describes this second technique as follows:

> he identified social formations and social types, tracing their milieux by economic analysis and their modes of observation and writing by cultural analysis. (Williams, 1973b)

This technique includes the now famous identification of the social types of 'the *flâneur*' and 'the ragpicker' which, while initially introduced as a form of background to the content of Baudelaire's lyric poetry, are also employed immanently to account for Baudelaire's mode of aesthetic observation and composition. While not employing social types as such, Williams clearly sets out to elaborate a similar thesis: 'that Dickens could write a new kind of novel...because he shared with the new urban popular culture certain decisive experiences and responses' (*TEN*, p. 28). This new urban popular culture is not presented by Williams with the micrological specificity with which Benjamin celebrates, for instance, Dickens's reminiscence of the gas lamps of Genoa.[61] Yet Benjamin's account of the rise of the commercialized *feuilleton* and the literary marketplace clearly shares much with the social histories of cultural institutions in *The Long Revolution*. Moreover, if more common ground were needed, Benjamin bookends the whole of the second draft with allusions to and citations from Marx's *Brumaire*, and develops from that text his analysis of examples of what

Williams would call formations, the professional *conspirateurs* and *la bohème*.[62] Moroever, Benjamin's attempt to locate both the *flâneur* and Baudelaire's literary stance within a process of poetic self-recognition of the proletarianization of the petit bourgeois class (to which Baudelaire belonged) more than slightly echoes the 'literary representative' passage of *The Brumaire*.[63]

It is hardly surprising then that Williams drew heavily on this Benjamin text in his assessment of the categories of correspondence and homology in *Marxism and Literature*. Williams turns to this material to further develop his interest in the Frankfurt School's 'positive mediation'. He chooses to focus on Adorno's critique of Benjamin's 'revised' conception of 'dialectical images' in the 1935 *exposé* ('Paris, Capital of the Nineteenth Century').[64] The section to which Adorno's comment refers occurs in the first section, 'Fourier, or the Arcades', where Benjamin introduces his broad thesis concerning the arcades and, in particular here, their role as inspiration for Fourier's utopian vision of a phalanstery. Within this account Benjamin provides this pivotal 'definition' of dialectical images:

> To the form of the new means of production, which to begin with is still dominated by the old (Marx), there correspond images in the collective consciousness in which the new and the old are intermingled. These images are ideals, and in them the collective seeks not only to transfigure, but also to transcend, the immaturity of the social product and the deficiencies of the social order of production. In these ideals there also emerges a vigorous aspiration to break with what is out-dated – which means, however, with the most recent past. These tendencies turn the fantasy, which gains its initial stimulus from the new, back upon the primal past. In the dream in which every epoch sees in images the epoch which is to succeed it, the latter appears coupled with elements of pre-history – that is to say of a classless society. The experiences of this society, which have their store-place in the collective unconscious, interact with the new to give birth to the utopias which leave their traces in a thousand configurations of life, from permanent buildings to ephemeral fashions. (Benjamin, 1973, p. 159)

Williams deliberately sets aside any linkage by Benjamin of dialectical images with 'the familiar abstractions' of myth, a collective unconscious or subjectivist conceptions of creativity. Rather, he prefers Benjamin's interest in '"the historical process", in particular in relation to his awareness of the changing social and material conditions of different kinds of actual art-work' (*M&L*, p. 103). Hence Williams states that the idea of dialectical images 'needs definition' and, rather than cite the passage above, relies on Adorno's critique thereof:

> Adorno complained that, in Benjamin's hands, they were often in effect "reflections of social reality" reduced to "simple facticity". "Dialectical images", he went on to argue, "are models not of social products but rather objective constellations in which the social condition represents itself". They can "never be expected to be an ideological or in general a 'social product' ". This argument depends on a distinction between "the real social process" and the various fixed forms, in "ideology" or "social products", which merely appear to represent or express it. The real social process is always mediated, and one of the positive forms of such mediation is the genuine "dialectical image". (*M&L*, p. 103)

As he so often does, Adorno here distinguishes ideological products 'worthy' of emancipatory critique from other social phenomena.[65] Benjamin's dialectical images clearly fail this criterion for him. Once again he is alluding to the need for a dialectical mode of criticism with immanent and transcendent moments. Williams's commentary on Adorno here is ambiguous. He cautiously endorses – or, on a different possible reading, reconstructs to his own satisfaction – Adorno's deployment of the category of mediation as a 'positive' one that recognizes the constitutive role of 'the medium'. Yet he also endorses Benjamin's technique of seeking correspondences by 'lay[ing] one process beside another... to explore their relations'.[66] However, as Williams's subsequent discussion makes obvious, this practice refers to the procedures Williams endorsed in his initial 1973 review of Benjamin rather than the above definition of 'dialectical images'. This would appear to be the significance of Williams's moving from Adorno's critique rather than Benjamin's 'definition' above. At the very least, for all his emphasis on the micrological dimensions of cultural practices, Benjamin's reiteration of their 'epochal' character would have been enough to arouse Williams's suspicion. Moreover, there is little evidence that Williams held similar interest in any such fragmentary cultural objects.

Rather, Williams 'recentres' the discussion upon Adorno's and Benjamin's contribution to an understanding of the categories of homology and correspondence. He presents this methodological overview as a typology (*M&L*, pp. 104–5). He adduces three types of homologically correspondent analysis summarized in Table 3.2.

The techniques attributed to Benjamin undoubtedly refer to 'The Paris of the Second Empire in Baudelaire'. Williams concludes more generally:

> A cultural phenomenon acquires its full significance only when it is seen as a form of (known or knowable) general social process or structure. The distinction between process and structure is then crucial. Resemblances and analogies between different specific practices are usually relations *within* a process, working inwards from a particular form to a general

Table 3.2 Benjamin and Adorno: correspondence and homology

Mode of correspondence	Example in Benjamin/ Adorno (on Williams's reading)	Example in Williams[a]
Resemblances between seemingly different cultural practices that are responses to 'a general social process' All evidence adduced highly specific but very extensive across different practices	Benjamin: Baudelaire's new poetic methods and (analysis of) ragpickers and bohemians	resemblances established between developments in popular press and theatre[b] resemblances established between cinema, popular culture and other arts and general cultural movement recognizing 'profound perceptual shifts'[c]
Analogies established between different social and literary forms Evidence direct and specific Analysis formal but correspondence is of literary stance	Benjamin: analogy established between the flâneur and corresponding forms of mobile and detached journalistic observation, and literary stance across literary and social forms	Dickens's adoption of popular cultural elements 'of the street' (*TEN*)
Displaced connections (homologous structures) Evidence direct and specific Analysis not only formal but also consists of consequent deduction	Adorno: negative relation between Viennese 'number games', tonal music and backward state of Austrian social development[d]	Contradiction between mobile privatization and suburbanization 'resolved' by social form of broadcasting (Section 6.5) The 'Bloomsbury Fraction'

[a] These parallels are postulated by myself, not Williams.
[b] 'The Press and Popular Culture: an historical perspective' (Williams, 1978b).
[c] 'British Film History: new perspectives' (Williams, 1983b).
[d] Adorno (1973b, p. 62, fn. 24). Remarkably, Adorno employs the production paradigm in this example, but again, Williams fails to comment on the parallel with his own work.

form. Displaced connections, and the important idea of *homologous structures*, depend less on an immediately observable process than on an effectively completed historical and social structural analysis, in which a general form has become apparent, and specific instances of this form can be discovered, not so much or even at all in content, but in specific and autonomous but finally related forms.

These distinctions have considerable practical importance. Both "correspondence" and "homology", in certain senses, can be modes of exploration

> and analysis of a social process which is grasped from the beginning as a complex of specific but related activities. Selection is evidently involved, but as a matter of principle there is no *a priori* distinction between the necessary and the contingent, the "social" and the "cultural", the "base" and the "superstructure". Correspondence and homology are then not formal but specific relations: examples of real social relationships, in their variable practice, which have common forms of origin. (*M&L*, pp. 105–6)

As is his common practice, Williams is here arguing on two fronts simultaneously. On this occasion, however, the two 'fronts' are already in dialogue. While Williams is keen to assert the role of art as 'a primary process', he is equally attentive to the needs of 'an effectively completed historical and social structural analysis' where the 'general form has become apparent'. This 'form' clearly predominates here over the crystallizations Benjamin addressed as 'dialectical images'.

However, having disposed of those elements of 'totality' that he finds unsatisfactory, Williams holds an advantage over many other commentators on the Adorno/Benjamin dialogue. For the differences between these methods can be accommodated within Williams's mode of hegemonic analysis. The techniques identified with Benjamin are processual and thus suited to emergent forms and 'fully apparent' cultural forms, Adorno's displaced connections more suited to the location of the analysis of 'fully apparent' forms within a determinate society. Table 3.2 therefore speaks directly to Table 3.1. Indeed, Williams closes this chapter of *Marxism and Literature* by reiterating his critique of Goldmann and introducing the next chapter on hegemony as a solution to Goldmann's 'epochal' failings.

Thus Williams's embrace of 'homologous structures' is influenced not only by his reading of Goldmann but quite crucially by his reading of Adorno and Benjamin. Moreover, his careful selection from the range of methods he believes are practised by Adorno and Benjamin reveals that his developing mode of conjunctural analysis was anything but 'culturalist'. Williams clearly saw a need to complement immanent cultural analysis with the analysis of a 'general social process'. He had already experimented with this technique in *Television* three years earlier.[67]

This downplaying of the role of 'content' for 'form' in the discussion of displaced connections in the citation above also raises interesting questions.

3.4 Ideology, critique and form

I have called this chapter 'From criticism to critique'. So in what sense does Williams move to something like Adorno's conception of 'emancipatory (ideology) critique'? First, it is clear that Williams never embraced that term as such nor even included it within his historical semantic surveys of the categories of ideology and criticism. Rather, as I argued in Chapter 1, Williams practises a form of immanent critique of argumentative prose in *Culture and Society* that grows out of his subversion of practical criticism. However, he does not reflect extensively on this practice. Yet, as we have just seen, he comes very close to – or effectively succeeds in – recognizing this practice in Adorno.

Before moving on to assess briefly Williams's own analytic practice in this context, the question of ideology is worthy of further attention. Certainly, the concept has a vexed status in Williams's work but this has been overestimated by commentators. In his longest discussion, in *Marxism and Literature*, he asserts that there are three chief meanings within the Marxian tradition:

(i) a system of beliefs characteristic of a particular class or group
(ii) a system of illusory beliefs – false ideas or false consciousness – which can be contrasted with true or scientific knowledge
(iii) the general process of the production of meanings and ideas (*M&L*, p. 55).

In his detailed discussion of Marx and Engels, however, Williams correctly deduces that for them (in *The German Ideology*) the concept – arguably in all three senses above – is *further* confined in its reference to the products of the professional 'ideologists', intellectuals who consciously or not, develop systems of thought which legitimate an existing social order by mechanisms such as 'eternalizing' and 'naturalizing' perspectives and assumptions that are actually historically determinate.[68] This position is compatible with both '*The Brumaire* solution' (as in meaning [i] above) and its 'structural limitation' and, further, with Adorno's conception of (emancipatory) ideology. Adorno would further delimit the range to elaborated theories and major works of art. Even then the process of critique, for both Marx and Adorno, allows for a redemptive potential in both elaborated ideologies and art.

Crucially, this position is not compatible with another sense of ideology that was pervasive in academic discussion at the time Williams was writing *Marxism and Literature*. This sense would expand this conception to *all* forms of 'consciousness', even, especially, the everyday consciousness of the non-intellectual. Certainly this conception was strongly influenced by Althusser's 'structuralist' interpretation of Marx. For Stuart Hall, for example, all consciousness is necessarily 'decentred' (Hall, 1977b, p. 320) and so ideology came to mean for him:

> the mental frameworks – the languages, the concepts, categories, imagery of thought, and the systems of representation – which different classes and social groups deploy in order to make sense of, define, figure out and render intelligible the way society works.

> The problem of ideology, therefore, concerns the ways in which ideas of different kinds grip the minds of masses, and thereby become a 'material force'. (Hall, 1983a, p. 59)

Two issues here become conflated by Hall, the social production of ideologies and the socio-political consequences of the 'naturalizing' legitimative role that such ideologies go on to play. Hall's account here, like Althusser's, leaves no theoretical space for the social production of ideologies. This was almost a non-issue for Althusserians as all such practices were seen to take place 'within' 'ideology in general'. As recently as 1997, Hall stated that he accepted 'the Althusserian argument about the impossibility of getting outside of ideology' (Hall, 1997, p. 30).

Much – including the pertinence or otherwise of many 'post-Marxist' rejections of the very use of the concept of ideology – turns on these apparently arcane distinctions.[69] Williams is surprisingly explicit about the necessity of recognizing the distinction between the social production of ideologies and their role as means of legitimation, as early as the 'Base and Superstructure' essay. Shortly before he introduces the concept of hegemony but *after* he has argued for the revision of all elements of the base and superstructure metaphor, he mounts a limited *defence* of the maintenance of 'the superstructure' in the societal use of the metaphor:

> I have great difficulty in seeing processes of art and thought as superstructural in the sense of the formula as it is commonly used. But in many areas of social and political thought – certain kinds of ratifying theory, certain kinds of law, certain kinds of institution, which after all in Marx's original formulations were very much part of the superstructure – in all that kind of social apparatus, and in a decisive area of political activity and construction, if we fail to see a superstructural element we fail to recognize reality at all. These laws, constitutions, theories, ideologies, which are claimed as natural, or as having universal validity or significance, simply have to be seen as expressing and ratifying the domination of a particular class. (1973a, p. 7; *PMC*, pp. 36–7)

The key to this statement is, I would suggest, Williams's use elsewhere of the term 'processual'. As we saw in the previous section, Williams is keen to distinguish between processual and 'fully apparent' cultural phenomena. It is this processual dimension that Williams holds up against a position like Hall's. Thus 'processes of thought' are not superstructural while 'ratifying theories' are. That is, such 'processes' are deemed to be 'in play' until it can be demonstrated that they perform an ideological function of ratifying an existent order. From at least as early as *Modern Tragedy*, Williams employed an understanding of ideology largely delimited to such legitimative practices.

However, even this legitimative ideology or cultural form remains susceptible to his immanent critique and its determinate location within a contingent hegemony.

Williams later found a Marxian warrant for this processual position in Marx's and Engels's grounding of their discussion of ideology in *The German Ideology* in a conception of an apparently pre-ideological 'practical consciousness' prior to the development of a mental/manual labour division. As Williams deploys this more fully in his discussion of language, it will be set aside to Section 4.2. But such a conception was, for Williams, evidently compatible with his rebuilding of Goldmann's genetic structuralism from its subcategories of 'consciousness'.

Williams's had actively deployed his legitimating conception of ideology in works such as *Modern Tragedy* (1966) and *The Country and the City* (1973). *Modern Tragedy* is an unusually hybrid work with a first section consisting of a long assessment of 'argumentative prose', from literary criticism to political theory, and a second which discusses overtly literary materials.[70] As noted in Chapter 1, the assessment of 'the tradition' of 'tragic ideas' resembles that of 'culture' in *Culture and Society*.[71] Even more than in that text, however, *Modern Tragedy* employs a mode of immanent emancipatory ideology critique. We can say *ideology* critique here because this time Williams explicitly categorizes 'the tradition' as ideological because of the legitimative role it came to play.[72]

However, there is an interesting dynamic to this legitimation process that links Williams's practice of ideology critique with his model of hegemony. Williams challenges the forms of 'universalist' meanings attributed to tragedy by the critical tradition. He argues that the location of tragedies within moments of social crisis has there been largely repressed. Contemporary tragic theory is thus incapable of recognizing – and even denies – the tragic dimensions of modernity. This misrecognition is related to contemporary ideas of order and disorder. Accordingly Williams provides an assessment of the fate of liberalism and Romanticism in this context, arguing that their immanently revolutionary potential has been overcome by the reworking of their universalist goals into a utilitarian ideology of 'modernization' and nihilism respectively. *These* ideologies have so come to play a legitimative role. They leave no space for a recognition of 'the structure of tragedy within our own culture' that is (in part) the relationship between suffering and the struggle for social change in moments of crisis.

However, Williams's analysis also stresses the contingency of the subordination of 'the idea and theory' of tragedy to the pressures 'of contemporary ideology and experience' (*MT1*, p. 45). Accordingly it is '...necessary to break the theory if we are to value the art' (*MT1*, p. 46). 'Value' here means to bring a tragic interpretative perspective to modern drama but it is also the term Williams uses to identify a similar failing within liberalism's mutation into a utilitarian ideology of modernization that legitimates the 'separation of change from value' (*MT1*, p. 73). 'Value' so refers to the repressed normative dimensions of both literary and political theory.

When Williams turns to 'value the art' in the plays that he regards as 'modern tragedies', he clearly is no longer dealing with 'argumentative prose' nor is he isolating 'the words on the page' of individual works as his practical-critical training had advocated. Rather he addresses *cultural forms*.[73] Four years later in his Goldmann lecture, Williams described what he attempted in this section of *Modern Tragedy*:

> I had become convinced in my own work that the most penetrating analysis would be of forms, specifically literary forms, where changes of viewpoint, changes of known and knowable relationships, changes of possible and actual resolutions, could be directly demonstrated, as forms of literary organization, and then, just because they involved more than individual solutions, could be reasonably related to a real social history, itself considered analytically in terms of basic relationship and failures and limits of relationship. This is what I attempted, for example, in *Modern Tragedy*... (1971, p. 13; *PMC*, p. 26)

We are partly returned in this passage, however, to the relative vagueness of Williams's earlier methodological language. He formulated this position with much greater clarity in his discussion of *The Country and the City* in *Politics and Letters*.

Williams's analysis of poems in celebration of English country houses in *The Country and the City* provided the working example for discussion in *Politics and Letters*, and can serve a similar purpose here. The former was published in 1973 and the period of its writing overlapped with at least that of the Goldmann lecture and possibly 'Base and Superstructure'.

In response to the interviewers' introduction of an understanding of cultural forms as means of production, Williams responds:

> My project, a very difficult one in which I am not sure I always succeeded, was quite different: it was to try to show simultaneously the literary convention and the historical relations to which they were a response – to see together the means of production and the conditions of the means of production. For the conditions of the means of production are quite crucial to any understanding of the means of production themselves. (*P&L*, p. 304)

Williams contrasts this position with an ahistorical technicist *formalism* which might also speak of production by cultural forms (and so regard them as means of production) but not recognize the social conditions of production of those forms. The key link for Williams is the recognition that the conventions that become embedded in cultural forms (and linguistic conventions) are the product of social relations susceptible to social and historical analysis. It is at this point that the Adorno/Benjamin typology of modes of correspondence (Table 3.2) might be brought into play.[74]

Williams's analysis in *The Country and the City* accordingly moves initially from an immanent analysis of specific poems. Three by Ben Jonson (*Penshurst* and *To Sir Robert Wroth*) and Carew (*To Saxham*) are asssessed against a criterion of 'truth' derived immanently from within another 'tradition', that is, from Crabbe's rejection of the 'neo-classical pastoral' tradition in his *The Village* (1783). Jonson's and Carew's country house poems are presented as the final development in a process of transformation that began in the late seventeenth and early eighteenth centuries from the 'conventional pastoral' form into 'what can be offered as a description and thence an idealization of actual English country life and its social and economic relations' (*C&C*, p. 38). This is the 'falsehood' challenged by Crabbe. Only in the context of this convention does Williams assert the significance of these poems' suppression of references to those whose labour produced the cornucopia of the patron's dining table celebrated by Jonson and Carew.[75] The crucial shift Williams describes is a move from a complaint about the present from the perspective of a mythologized past in early pastoral, to one of celebratory 'naturalization' of the social order of the present.

It is this sense of 'naturalization' that Williams categorizes as 'ideology', specifically an 'open ideology of improvement' which legitimates the enclosures of small holdings by the 'rising' class of agrarian capitalists (*C&C*, p. 80).

In the discussion with his interviewers, Williams insists that it is the selective tradition regarding the ideological innocence of the *form* of the country house poem that he is challenging. There was in effect a complicity between orthodox literary criticism and the conventions of the poems. Williams thus elaborates the two 'foolproof' stages to his mode of critique:

> The first is that the very process of restoring produced literature to its conditions of production reveals that conventions have social roots, that they are not simply formal devices of writing. The second is that historical identification of a convention is not a mere neutral registration, which is incompatible with judging it. Indeed, as literary evaluation proper is concerned, I would say that while there is a not unhelpful mode – I wouldn't put it stronger than that – of distinguishing between good and bad examples within a convention, the crucial evaluative function is the judgement of conventions themselves, from a deliberate and declared position of interest... *You have to be able to go beyond an understanding that the poems are not records of country-house experience, to the realization that these conventions produce actions and relationships, as well as poems, and as such they stand to be judged.* It is not difficult to distinguish between poems by Jonson and Carew – the former are better written in a perfectly normal sense than the latter. But what is more important than that

> distinction is the distinction of the convention: the capacity to see what the form was and what it was producing. Certain conventions do less than others. If there is still place for evaluation in literature, then that is what has to be valued. This is not the same as saying, although one also says, that the poems are not like history. For a convention could resemble no actual history at all, yet be positively productive by its representation of possible situations. The soundest conventions are not always realist, although that is more often the case than not. Each convention must be assessed by what it is rooted in and what it does: an assessment that is related to a much more general historical judgement that is also an affiliation – not history as all that has happened but where oneself is in it. (*P&L*, pp. 306–7; emphasis added)

Here a clear difference does emerge between Williams's method and those of Goldmann and Adorno. Goldmann and Adorno would eliminate the 'poorly written' examples from consideration as they tend to presume a harmony between 'coherence' and the goals of their immanent critiques. It is this stronger, because less 'discriminating', (social) formalism in Williams that also keeps his analyses closer to popular forms. Here Williams is interested, also if not equally, in the cruder examples of the convention that render its legitimative functions more evident. As we saw in Section 1.4, this was in effect the rationale for the 'documentary' conception of culture in *The Long Revolution*. It is in such instances that Williams demonstrates his preference for prioritization of processual form (or 'practice') over 'works'. Nonetheless, he also reserves space for those exceptional works which do so much more than merely reproduce the convention, so that returning these works to their conditions of composition would not be 'a full account of their composition' (*P&L*, p. 328).

Yet when his interviewers pursue him further on a particular 'judgement' he makes in *The Country and the City* of (orthodox) Marxism's 'simultaneous damnation and idealisation' of capitalism's productive forces, an exchange takes place which again evokes a dialectical model of critique from Williams:

> *NLR*: . . . The Marxist tradition does insist that capitalism, feudalism, or slavery in the ancient world, all represented massive structures of class oppression and exploitation, yet that each was also empirically related to forms of greater human emancipation. Were you really rejecting that?
>
> *RW*: No. Let me give you an example where I have taken precisely that position, and been attacked for it on the left. I have emphasized that the achievement of the bourgeoisie in the creation of the modern press was a major historical break-through. I have no hesitation at all about declaring that. The advent of the bourgeois newspapers was an absolute historical

progress, which one must acknowledge even as an absolute opponent of the contemporary bourgeois press. I don't find any difficulty in making that kind of judgement. I wouldn't see it as reasonable to criticize the late 18th- or early 19th-century press in England because it was bourgeois.

NLR: Well, to take your criterion for Jonson, it didn't exactly report the life of the working class.

RW: I wouldn't limit the judgement to that. The emergence of this press was progressive. I quote it precisely to show that I am wholly in sympathy with reasonable uses of damn this/praise this. For by the mid-19th century the bourgeois press was consciously attempting to squeeze out, buy out, outsell, outcapitalize the popular radical press. By then, even if it was expanding certain areas of bourgeois liberty, it was a negative force. I think there is a scientific mode of attention in which damn this/praise this is right, but there is a mode of conventional inattention where it is profoundly wrong. That inattention is often related to a confidence that was very typical of the Communist parties when I was young, that you could damn-and-praise at will because you knew what the next epoch of human history would bring. (*P&L*, pp. 310–11)

This exchange has several interesting implications. It demonstrates how readily Williams regarded the critique of 'literature' and 'the press' as comparable if not equivalent. In effect, they qualify for an analogical correspondence as in Table 3.2. While moving no further than he had already done in *The Long Revolution*, his 'object of analysis' is broadened well beyond the territory of the abandoned practice of literary criticism. Williams had included a history of the British press as one of the 'social history' chapters of *The Long Revolution* and had maintained an interest in the area.[76] But in 1978 he had published two new articles on press history and policy which likely informed this comment.[77]

The passage also provides an interesting example of the positive dimensions of 'the productive capacity of bourgeois society, or its political institutions' that Williams mentions earlier in the same exchange (*P&L*, p. 307). Here, more clearly than the vexed case of literature, an immanent criterion of judgement readily emerged, albeit qualified by his 'absolute opposition' to 'the contemporary bourgeois press': the once revolutionary liberal goal of democratic citizenship. Here too we can see an evident continuity with the normative goals of the early Williams discussed in Chapter 1 and a prefiguration of the parallels to be drawn with Habermas's public sphere thesis.[78]

Yet again this raises the comparison with emancipatory ideology critique. The movement of argument in the above passage is uncannily similar to that elaborated in Williams's 1969 essay on Marcuse.[79] The bourgeois press, like 'culture', could be recognized for its initial emancipatory possibilities but has since ossified – contingently – into an affirmative force. Neither assessment for Williams is arbitrary or irreversible. Rather, immanent analysis must be accompanied by historical determination of the 'limits' of this progressive dimension of the ideology of the rising or dominant bourgeoisie.[80] This in turn activates his earlier comment above that other 'actions' might flow from the same conventions that shaped literary forms. It is these actions that would require counter-hegemonic contestation. The complexity of Williams's mature position thus starts to become more evident.

We so have sufficient evidence to assert that rather than practise a 'literary sociology' – a central element of the charge of 'culturalism' examined in Section 1.3 – Williams instead practises a form of emancipatory critique. I would suggest that this technique pervades all Williams's major 'cultural' critiques, from the historical semantic analysis of particular 'keywords' through much of his literary 'criticism', and on to his recovery of the emancipatory potential within superficially unpromising aesthetic and theoretical, and even some popular cultural forms. Moreover, there is a strong 'social formalist' dimension to this procedure that requires further elaboration in the next chapter.

Such an expanded conception of critique is a more appropriately dialectical mode of application of Williams's dialectically 'expanded' conception of 'culture' than the reductivist usage of 'whole way of life' which has dominated such discussion in cultural studies and beyond. That is, the dialectically expansive conception of culture should be seen as enabling an expansive mode of immanent socio-cultural critique rather than merely an expanded 'object' of analysis devoid of a corresponding 'method'.

The following chapters aim to elucidate the major dimensions of Williams's later writings on the implications of this achieved practice.

4
Social Formalism

4.1 Against formalism and 'the language paradigm'

Chapters 2 and 3 introduced Williams's production paradigm and developments within his cultural materialism and within his mode of critique. But Williams also develops another position as he responds to the developing influence of structuralism within and beyond cultural studies. That position he both claims as his 'own' and identifies as a 'fellow travelling' one he recognizes within others. He calls that preferred position *social formalism*.

By 1976 there had appeared in Britain a number of projects with Louis Althusser's 'structuralist Marxism' as a common thread. Not only had Eagleton's *Criticism and Ideology* been published, the *Screen* project's innovative explorations of the relationship between 'the classic realist text' and semiotics were well underway, and the fame of the Birmingham CCCS – based heavily in its own appropriation of structuralist and semiotic methods – was growing. In his first major criticism of 'the structuralist version of Marxism', Williams characterized it as:

> especially strongly established in anthropology and linguistics. This tendency has achieved an important critique of earlier ideas of *superstructure*, and an equally important rethinking of the concepts of *structure* and of *practice*. But, more than any other, it is a theoretical displacement of real cultural practice, in the interest of what is, at the level of inquiry, a technology. Its preoccupation with formalized structures, and with systematic determinations, is in sharp contrast . . . with earlier concepts of reflected or reproduced content and of a centrally determined system (*base and superstructure*). But in just this preoccupation it recapitulates, in new technical forms, an objective idealism which has indeed always been, in cultural analysis, an attractive position. The reductionism inherent in older kinds of Marxist analysis – a reduction of specific content to other content – has been superseded and then replaced by a new reductionism, in which the privileged observer reduces

> all practices to systematic configurations, which alone create and contend. (1976b, p. 503)

The reference to 'linguistics and anthropology' is almost certainly an allusion to the dependence of the Althusserian project on Lévi-Strauss's structuralism. That dependence had been strongly argued in well-known critiques of Althusser's work.[1] The curious reference to a technology alludes to the instrumental *technicism* of the formalism that Williams wishes to reject.[2] But the reference to a 'privileged observer' was perhaps of most importance to Williams.

The related charge of 'objective idealism' is more precisely formulated, as we shall see, as one of converting 'all social practices into forms'.[3] The practice/form distinction is, as we saw in previous chapters, part of Williams's own processual emphasis developed in his critical reconstruction of Goldmann. As Williams insisted in his 1974 inaugural professorial lecture, he did not 'turn to sociology' but, rather, had always regarded sociality as embedded in cultural forms.[4] However, as Williams developed his own theoretical position he also made it plain that not all sociality could be discovered by formal analysis. The social formal analysis Williams wishes to recommend decidedly does *not* convert 'all social practices into forms'. As we shall see in Section 4.4, he charged the Birmingham CCCS with that failing and with 'privileged observation'.

The material drawn on throughout this chapter ranges from Williams's earliest criticisms of 'synchronic structuralism' in 1976 to his 1986 lecture, 'The Uses of Cultural Theory'.[5] His basic position changed little in those ten years. In 'The Uses of Cultural Theory', he explicitly identifies 'the language paradigm' as his chief point of disagreement with contemporary 'cultural theory'.[6]

By 'the language paradigm' Williams means what is more often called the linguistic paradigm or 'linguistic turn': the body of work (especially that of Lévi-Strauss) that claimed that non-linguistic phenomena were 'structured like a language'. Three relevant potential lines of development follow from this claim. For the 'structuralist' consequences of the linguistic paradigm in 'other' disciplines, the key lineage of relevance is Roman Jakobson's 'properly phonemic' (re)orientation of Saussure's linguistics.[7] Lévi-Strauss's initial goal was 'the transposition of the phonemic method to the anthropological study of primitive peoples' (1963, p. 35) (briefly introduced in Section 3.2).[8] Second, there is the distinct but related project of semiology usually associated with Roland Barthes's early work, and which also built directly upon Saussure.[9] Both these projects deliberately sought to expand their fields of application beyond the 'materiel' of language.

However, while Williams's own work also pointed beyond language *per se*, most of his analyses remained strongly tied to the 'materiel' of language. Accordingly, he also necessarily recognized a third line of interest in 'structuralist poetics', the consequences of the 'linguistic paradigm' for 'literary criticism'.[10] It is here that the Russian formalist project assumed a greater significance, and it is chiefly in this context that Williams identified a social formalist alternative.

The Russian formalists' principal task was the isolation and investigation of the formal properties of a specifically literary phenomenon – the literary device that produced 'literariness'. Bakhtin and Medvedev's well-established connection between Russian formalism and the Russian futurist literary avant-garde appears to have inspired Williams to develop his formational model to account for the receptivity of contemporary theoretical configurations with an 'avant-gardist' orientation.[11] As we shall see, however, this linkage begins with his earlier critique of McLuhan as a representative of the 'later stages of the formalist tradition'.[12]

Williams introduces the distinction between formalism and social formalism in *Marxism and Literature*, and develops it further in *The Sociology of Culture* and *The Politics of Modernism*. The later discussions imply that by 'social formalism' he meant exclusively the work of the 'Vitebsk group', those formalists who criticized the main Russian formalists during the late 1920s, a group including Bakhtin, Medvedev and Vološinov. However, in *Marxism and Literature* it is implied that the work of the Prague Circle led by Roman Jakobson – or at least that of Jakobson's colleague, Jan Mukařovský – is also social formalist. In a revision to the entry on 'formalism' in the second edition of *Keywords*, Mukařovský is explicitly grouped with Vološinov as a social formalist.[13]

The significance of these emphases is considerable. As Doložel has pointed out, 1970s Western textbook accounts of structuralism – on which much English language reception of that project was based – tended to share a particular narrative of its development. Saussure's initiatives in structural linguistics and the work of the Russian formalists were regarded as important precursors, but 'structuralist poetics' *per se* was regarded as foundationally French. In an anticipation of future narratives of the formation of cultural studies, what was influential in Paris became the criterion of selective historical emphasis in these 1970s accounts. Mukařovský's 1946 Paris lecture on Prague structuralism went unnoticed, as did Goldmann's attempts to shift the terrain of discussion of the Parisian structuralists twenty years later.[14] Thus were crucial innovations of Prague structuralism sidelined.[15]

Clearly, Williams was drawing on all possible sources from which he might develop an alternative to the technicist formalism he saw within 'French structuralism'. In so doing he was at times considerably in advance of much contemporary English language scholarship.[16] The social formalism also grew from the emphases Williams placed on form in his reading of Western Marxists other than Gramsci, most obviously Goldmann. As early as 1976 Williams situated the recovery and reconstruction of those (social) formalist elements of 'Western Marxism' and his own work on dramatic forms, as 'points of entry into a sociological analysis'. He so provided a preliminary contrast of 'such work' (Lukács, Adorno, Benjamin and his own *Drama From Ibsen to Brecht*) with a technicist formalist tradition thus:

> All such work overlaps, of course, with a quite distinct and often antagonistic tendency in the analysis of art, which can be traced through modern European culture in its stages of formalism, practical criticism, new criticism and synchronic structuralism. Work on form, in its widest sense, in these other tendencies, has been of the greatest importance, but at significant moments in each phase it has become explicitly anti-sociological, postulating separable or at least radically distinct areas of practice, and using work on tradition in the strictest and most formal ways. (1976b, p. 502)

By implication, then, we have here already a broad definition of 'social formalism': a position that recognizes the significance of cultural forms as more than mere vehicles of social determinacy (as vulgar Marxism would have it), but which nonetheless requires that they provide 'a point of entry into a sociological analysis'.

But a point of entry is not the same as a sociological analysis. Let us start then with Williams's challenge to the most basic foundations of 'the language paradigm'.

4.2 Language, signification, practical consciousness

Williams's social formalist project commences in his chapter on language in *Marxism and Literature* where he embraces Vološinov's critique of Saussure as 'abstract objectivism'. Vološinov's 1930 work, *Marxism and the Philosophy of Language*, had been translated into English in 1973.[17]

The features of Saussure's argument that Williams, largely following Vološinov, finds 'abstract objectivist' are the following:

(a) Saussure's 'Principle 1', the arbitrary character of the sign, that is, Saussure's most basic premiss that rejects a 'naming' conception of the relationship between words and meanings. In order to facilitate his break with this 'commonsensical' presumption, Saussure uses the concepts of (linguistic) sign and its (analytic) components, signifier and signified. The signifier is the auditory means ('sound-image') of the sign, while the signified is the meaning ('concept') it carries. In asserting the sign's 'arbitrary' character, Saussure means that the signifier/signified bond is not 'natural' and that, in principle, any meaning can be conveyed by any auditory means.[18]

Williams accepts the premise that no 'natural' bond exists, but he insists that the use of the term 'arbitrary' to characterize this non-correspondence conceals *the social conventions* by which the signifier/signified relation is 'fused' (*M&L*, p. 27).

(b) The ahistorical consequences of the langue/parole distinction. This is one of several binary dichotomies Saussure usually represents visually by axes. Their chief relevance for the author of *Keywords* is the question of changes in meaning that Saussure recognizes with his 'mutability of the sign'. Langue is the 'system' constituted by the conventions of linguistic practice that are *unreflectively* 'known' by each speaker and 'confronted as a state' (Saussure, 1966, p. 9). 'Parole' is the manifest speech that results from the operations of these conventions. Diachrony, in this context, is the horizontal axis of incremental changes; synchrony, the vertical axis of overarching paradigmatic rules within which change may occur. But the 'state' of achieved conventions is extremely conservative, despite the 'facts' Saussure acknowledges of frequent changes in signifier/signified relations.[19] Moreover, Williams rejects what he understands to be the psychologistic underpinning of this aspect of Saussure's work, as it turns on yet another reified model of 'individual/society'. He also criticizes subsequent structuralist theories that either privilege synchrony over diachrony or work with an impoverished conception of the diachronic. Both Williams and Vološinov note the parallel here with the objectivist potential in the pursuit of 'social facts' in Durkheimian sociology.[20]

For Williams the abstract objectivist moment arrives, not with the empirical classificatory methodology of comparative linguistics (with which he would have little disagreement), but rather when this is combined with the perspective of a privileged observer of *alien* material:

> in texts the records of a *past* history; in speech, the activity of an alien people in subordinate (colonialist) relations to the whole activity of the dominant people within which the observer gained his privilege... In a later phase of this contact between privileged observer and alien language material, in the special circumstances of North America where hundreds of native American (Amerindian) languages were in danger of dying out after the completion of European conquest and domination, the earlier philological procedures were, indeed, characteristically, found to be not objective enough. Assimilation of these even more alien languages to the categories of Indo-European philology – the natural reflex of cultural imperialism – was scientifically resisted and checked by necessary procedures which, assuming only the presence of an alien system, found ways of studying it in its own (intrinsic and structural) terms.... at the level of theory it was the final reinforcement of a concept of language as an (alien) objective system.
>
> Paradoxically, this approach had even deeper effect through one of the necessary corrections of procedure which followed from the new phase

> of contact with languages without texts. Earlier procedures had been determined by the fact that a language almost invariably presented itself in specific past texts: finished monologic utterances. Actual speech, even when it was available, was seen as *derived*, either historically into vernaculars, or practically into speech acts which were instances of the fundamental (textual) forms of the language... North American empirical linguistics reversed one part of this tendency, restoring the primacy of speech in the literal absence of "standard" or "classical" texts. Yet the objectivist character of the underlying general theory came to limit even this, by converting speech itself to a "text" – the characteristically persistent word in orthodox structural linguistics. Language came to be seen as a fixed, objective, and in these senses "given" system, which had theoretical and practical priority over what were described as "utterances" (later as "performance"). (*M&L*, pp. 26–7)

The asocial and ahistorical character of the Saussurean conception of langue is thus, for Williams, a product of this reified 'textualist' paradigm. The alienation of the observer from the 'observed' – and its colonialist underpinning – is clearly crucial to this development. At each of the elision points in the above citation, Williams makes what is broadly the same point: that this increasingly dominant linguistic paradigm prevented the development of an alternative which would have been premissed on a 'sense of language as actively and presently constitutive' (*M&L*, p. 26).

Unsurprisingly, Williams can trace this alternative perspective, as with his mature reworking of the Romantic legacy in the case of 'culture', to Herder.[21] He argues that this 'post-Romantic' legacy was never taken up and debated in linguistics on the scale of, say, the methodological dispute between hermeneutics and positivism elsewhere, and yet the bifurcation was even greater: between an absolute idealism and the hegemonic objectivism discussed above.[22] Williams nonetheless sees a potential in the over-idealized arguments of Herder, and especially Humboldt, concerning language as creative 'expression'. That potential is a conceptualization of language as an intersubjective activity embedded in social relations. Significantly for Williams, this Romantic line of thought occludes this intersubjectivity by withdrawing into a subjectivist psychologism.

For these reasons Vološinov's work is a major breakthrough in Williams's estimation. Unlike his Saussurean Russian formalist contemporaries, Vološinov actively embraced Humboldt's legacy of language as creative activity for his general linguistics.[23] But this is no naïve appropriation. Vološinov specifically positions his project as a critique of the legacies of *both* abstract objectivism and individualist 'expressivist' subjectivism.

Williams endorses Vološinov's challenge to the Saussurean conception of the sign, in particular Saussure's rejection of any notion that the signifier/signified bond might be immutable in social practice. Vološinov takes great pains to distinguish the socially dynamic existence of signs in contemporaneity from the 'semantic paleontology' of Saussurean and other manifestations of abstract objectivism.[24] Yet even Vološinov does not break with Saussure sufficiently strongly for Williams, as he continues to use the categories of 'sign' and 'sign-system'. Williams's sketched alternative would replace 'sign' with 'signifying element of language' (*M&L*, pp. 39, 42–3). Here the issue for Williams is less the division of the sign into the signifier/signified binary, than it is the psychologism of Saussure's *a priori* assumption that the signifier 'is not a material sound, a purely physical thing, but the psychological imprint of the sound' (Saussure, 1966, p. 65).[25]

Here Williams's cultural materialism comes into play. His alternative rests on shifting the process of signification beyond all such 'abstract' psychologization into the realm of 'practical material' activity. He follows Vološinov's 'objective psychology' by rendering the entire process of signification an objectivated one, as this passage demonstrates:

> A physical sound, like many other natural elements, may be made into a sign, but its distinction, Vološinov argued, is always evident: "a sign does not exist as part of a reality – it reflects and refracts another reality".[26] What distinguishes it as a sign, indeed what made it a sign, is in this sense a formal process: a specific articulation of a meaning. Formalist linguistics had emphasized this point but it had not discerned that the process of articulation is necessarily also a *material* process, and that the sign itself becomes part of a (socially created) physical and material world: "whether in sound, physical mass, colour, movement of the body or the like".[27] Signification, the social creation of meanings through the use of formal signs, is then a practical material activity; *it is indeed, literally, a means of production*. It is a specific form of that practical consciousness which is inseparable from all social material activity. It is not, as formalism would make it, and as the idealist theory of expression had from the beginning assumed, an operation of and within "consciousness", which then becomes a state or a process separate, *a priori*, from social material activity. It is, on the contrary, at once a distinctive material process – the making of *signs* – and, in the central quality of its distinctiveness as practical consciousness, is involved from the beginning in all other human social and material activity. (*M&L*, p. 38; emphasis added to central sentence)

While Williams's position here is consistent with the conception of production as objectivation discussed in Chapter 2, it so tends to attribute 'conscious design' to all uses of language.

Indeed, Williams even suggests that his resolution of this issue 'offers a basis also for a vital reconsideration of the problem of "subjectivity"' (*M&L*, p. 40). The 'price' of Vološinov's dual rejection of an abstract objectivism (as in the Saussurean system), and of an individualist subjectivism is the adoption of an 'enabling' psychology. That is, for Vološinov, 'from the point of view of content' the domains of ideology, signs and the psyche coincide (1973, p. 33). The enabling mechanism for Vološinov is the postulation of a dimension of 'inner speech' constituted by 'inner signs'. He separates this psyche from any identification with physiological processes. He admits that the status of this psyche is thus 'unclarified' (1973, p. 31), perhaps leaving room for revision of the hostile critique of 'Freudianism' published under his name three years before.[28] But this is a topography of the psyche not easily reconcilable with the Freudian unconscious.

Vološinov's conception of 'signal' holds particular interest for Williams. For Vološinov, a signal has no relationship with 'understanding'. Whereas signs always contain an 'ideological content', the signal is merely that which is recognized *unambiguously*, and 'relates to the world of technical devices, to instruments of production in the broad sense of the term' (as in a traffic signal) (Vološinov, 1973, p. 68).

Crucially, the signal thus has many of the properties Williams sees as problematic in the alienated post-Saussurean conception of the sign. The privileging of the synchronic systemic 'fixity' over the socially underspecified diachronic innovations leaves the Saussurean sign no more socially dynamic (for Williams) than the Vološinovian signal. He thus argues that what could be called the 'sociological deficit' of the Saussurean hypostasization can be attributed to 'a radical denial of practical consciousness' (*M&L*, p. 39).

Likewise, with some justification, Williams seizes on Vološinov's innovative conceptualization of the *multi-accentuality* of the sign, that characteristic by which it 'maintains its vitality and capacity for further development' (Vološinov, 1973, p. 23). Williams glosses Vološinov's multi-accentuality thus: 'It must have an effective nucleus of meaning but in practice it has a variable range, corresponding to the endless variety of meanings within which it is actively used' (*M&L*, p. 39).

We are thus provided with what is, in effect, a description of the project of *Keywords*. It is difficult to avoid the impression that Williams must have had a legitimation of his historical semantics in mind as he developed his advocacy of Vološinov. *Keywords*' historical semantics escapes Vološinov's charge of 'semantic paleontology' precisely because its criteria of selection privilege words that are still in active usage. To use the Saussurean terminology, they are signs whose signifiers have borne, and continue to bear, a variable range of signifieds.

Such an 'unfixing' of the Saussurean signifier/signified relation superficially resembles one of the more common perspectives in poststructuralist thought, best known via Derrida's *différance*: the potentially endless subversive 'play' of signification.[29] However, while Williams's critique does refuse to employ the usual Saussurean binary terms, it also *limits* the possible range of signifying 'play' by insisting on the socio-historical determinacy of the 'nucleus of meaning' as well as its shifts in any given conjuncture.

As Christopher Norris has recently argued of *Keywords*:

> it provides what structuralism couldn't and what post-structuralism couldn't and what certain currently fashionable 'post-analytic' or neo-pragmatist approaches to interpretation can't provide, that is, some way of explaining, not always with full clarity but often very suggestively, how it is that language both bears structures of consciousness and structures of feeling and at the same time articulates the changes that take place historically between them, and thus leaves room for the subject, that is, the conscious, intending, purposive speaker or writer. (Norris, 1997, p. 36)[30]

Williams's (promised) Vološinovian view of the role of subjectivity required for this dynamic process is significant. Speaking of the way in which a sign differs from a signal, he states:

> The true signifying element of language must from the beginning have a different capacity: to become an *inner sign*, part of an active practical consciousness. Thus in addition to its social and material existence between actual individuals, the sign is also part of a verbally constituted consciousness which allows individuals to use signs of their own initiative, whether in acts of social communication or in practices which, not being *manifestly* social, can be interpreted as personal or private. (*M&L*, p. 40)

Individual subjectivity is thus interestingly underplayed in this Vološinovian objective psychology. The sociality of language is deemed to be such that even the intimate use of linguistic signs relies on this verbally constituted (intersubjective) consciousness. As we shall see, this is quite consistent with Williams's view of the capacity of cultural forms to mediate the private/public boundary. Yet, consistent with Norris's commentary, this model also insists on the capacity of the subject to undertake initiatives within language. The social process socializes but there is no suggestion of the subject's being merely the means via which a linguistic system 'speaks' or is constituted 'in language'. Indeed, Williams insists in *Politics and Letters* on a pre-articulated level of consciousness, the pre-emergent, which enables initiatives within language and the use of linguistic *notations* as means of composition. Williams's interviewers point out that the Freudian conception of the unconscious is not necessarily that of a hard barrier between pure repression and pure self-consciousness, as he seems to assume. Rather, they say, 'the unconscious is an active structure which is at work in everything we do'. Williams's response is, characteristically, to typologize a range of possibilities, starting with the distinction between 'unwitting' and 'unconscious' (*P&L*, pp. 180–3).

Williams's alternative formulations to Saussure's tend to be modelled as means of production but this position is qualified. We saw in Section 2.5 that *notation* is the category by which Williams attempts to resolve the question of the relationship between objectivation and 'materialization'. Williams explicitly rejects a model of the work of art as text for that of notational practice. Likewise, notation is also the category Williams wishes to substitute for the role of 'sign' in the broader case of written language.[31] Here Williams draws this definitional line:

> To understand the materiality of language we have of course to distinguish between spoken words and written notations. This distinction, which the concept of "sign" fundamentally obscures, has to be related to a development in means of production. Spoken words are a process of human activity using only immediate, constitutive, physical resources. Written words, with their continuing but not necessarily direct relation to speech, are a form of material production, adapting non-human resources to a human end. (*M&L*, p. 169)

Clearly then, for Williams, the speaking of language is a processual activity reliant on bodily resources, while the linguistic notation is 'a form of material production' as it involves 'adapting non-human resources to a human end'. Here his model of language joins with his post-McLuhanist conception of means of cultural production. That is, Williams establishes this distinction on the basis of the means of cultural production deployed, his fundamental distinction being that between the body and the non-human resource of linguistic notations. Moreover, as in the object/notation distinction employed elsewhere, notations have the capacity for deployment within 'productive consumption' practices such as utterance. Williams's further discussion of notation implies a typological range of notational practices from the alphabetic to the more highly complex but this is not fully developed.

The use of 'activity' in the above passage is quite deliberate as it draws directly on the Romantic expressivist legacy. Humboldt's reformulation of language as creative activity is, as we saw, of crucial significance to Williams.[32] It provides him with one foil in his search for an anti-reductivist account of the relation between intersubjectivity and language. While Williams rejects Humboldt's idealism and his individualist subjectivism, he appears here to reconstitute the role of Humboldt's 'activity' to designate spoken language in contradistinction to the objectivated notations of written language.

On one occasion Williams passes the baton of further research to Ferruccio Rossi-Landi.[33] Rossi-Landi's attempt to extend the production paradigm to language is far more detailed than Williams's. However, he stops short of any assertion of a 'literal' application of the production paradigm. Rather, he restricts the relation to that of homological analogy between linguistic production and 'material production' (i.e. for Rossi-Landi, the production of objects).[34]

Humboldt explicitly *contrasts* his linguistic 'activity' with work (labour).[35] If we maintain Humboldt's distinction – that is that language-activity is *not* the same as labour – then we can also maintain the principle of non-reducibility that Williams equally wishes to assert by separating speaking from writing. Márkus argues that Marx's own presupposition of language-activity's sociality *provides a discrete account which contrasts with and stands separate from* the emphasis elsewhere in his work on the conscious character of labour-activity.[36]

Williams would draw the same normative distinction by leaving open the prospect that 'speaking' might not always be analogous to 'conscious' production. In effect, Williams's social formalism reverses the relationship between formalist linguistics and formalist poetics. Where formalist poetics draws heavily on formalist linguistics for its model of the text, Williams's social formalist linguistics is modelled on his alternative for the artistic 'text' (as object), notational practice.[37] This, in combination with the Vološinovian conception of signal, in turn informs his conception of genres/cultural forms.[38]

However, Williams also wishes to ground the emergence of language and other signifying practices in the intersubjectivity of 'practical consciousness', a category he derives from Marx.[39]

The source of Williams's conception of 'practical consciousness' is the discussion of language in Marx and Engels' *The German Ideology* (Marx and Engels, 1976):

> Only now, after having considered four moments, four aspects of the fundamental historical relationships, do we find that man also possesses 'consciousness'; but even so, not inherent, not "pure" consciousness. From the start the "spirit" is afflicted with the curse of being "burdened" with matter, which here makes its appearance in the form of agitated layers of air, sounds, in short, of language. Language is as old as consciousness, language is practical consciousness, as it exists for other men, and for that reason is really beginning to exist for me personally as well; for language, like consciousness, only arises from the need, the necessity of intercourse with other men. (Marx and Engels, 1976, pp. 43–4)[40]

It is not difficult to see the appeal for Williams of this formulation of 'practical consciousness'. The whole passage is, as he immediately comments, quite compatible with the Herderian Romantic emphasis on constitutive creative activity 'so far as it goes' (*M&L*, p. 29).

Williams correctly sees the above passage as a significant articulation of Marx's materialist conception of history.[41] Marx explicitly insists that the 'moments' be interpreted as 'aspects' rather than 'stages'. The first two of the four 'moments' to which Marx refers constitute a developmental dialectic of 'primary' need satisfaction and the concomitant positing of new needs. The third is one of Marx's rare references to the sexual division of labour – embedded (for him) in physical reproduction of the species. Like the reference to language, it receives little further elaboration in this text. The fourth aspect is an analysis of the first three and is an early articulation of the production paradigm itself. It posits that 'the production of life', through labour, thus operates within a dual relationship, natural and social. Crucially, as Williams recognizes, this passage points to 'simultaneity and totality' rather than the sequential causality of orthodox Marxism (or, we might add, of the nineteenth-century universal histories) (*M&L*, p. 29).

However, Marx's own immediate further elaboration of the reference to 'practical consciousness', in the continuation of the same paragraph as that cited above, is significant. As Rundell has put it, Marx provides an inventory of forms of consciousness so that 'consciousness embraces a continuum from myth to critique', which is quite definitely sequential (Rundell, 1987, p. 172). Williams appropriates 'practical consciousness' without any acknowledgement of this differentiation.[42] He frequently refers to this 'practical consciousness' as 'active', but his post-Romantic emphasis would hardly seem to include *The German Ideology*'s apparent inclusion of 'sheep-like or tribal consciousness' within the continuum of 'practical consciousness' (Marx and Engels, 1976, p. 44).

As Williams does note, this section of *The German Ideology* is 'part of their argument against pure directive consciousness' (*M&L*, p. 28). Accordingly, Marx's next step is to leave the discussion of simultaneous 'moments' to insist that the key historical development is the arrival of 'a division of material and mental labour': 'From this moment onwards consciousness *can* really flatter itself that it is consciousness of something other than consciousness of existing practice' (Marx and Engels, 1976, p. 45). This division of labour, of course, splits the very coupling Williams has embraced, 'practical consciousness'. Its significance for Williams's argument is considerable. Even allowing for the polemical character of the whole text, this preliminary definition of an ideologist points to Williams's need to reconcile at least this proposition with his own Vološinovian elaboration of 'practical consciousness'.[43]

Yet his 'processualism' and his embrace of *The Brumaire* solution did, in effect, provide Williams with a solution to this conundrum. Once such a modern division of labour exists, Williams could argue that the relationship between 'existent practice' and objectivated 'consciousness' would be a matter for determinate analysis of their respective 'limits and pressures'.

Williams's own use of 'practical consciousness' has a 'black box' role as heuristic sketch. The social process he gestures towards might more adequately be considered one of *disembedding*. Likewise, the pre-articulate consciousness he postulates is better considered a form of consciousness *embedded* within intersubjective social relations.[44] As we shall see shortly, Williams does employ a similar conception of disembedding in the case of cultural forms.

But what might be the determining status of 'structure' for this retrieved 'agency'? As we saw in Chapter 3, Williams embraced the concept of hegemony because it conveyed the 'deep saturation of the consciousness of a society' by a dominant culture. He also insisted that any determinate hegemony was contestable due to its dependence on the selective *incorporation* of meanings and values that are alternatives to the dominant. He thus set a task which closely anticipates that of his cultural materialist reflections on language: 'our hardest task theoretically, is to find a non-metaphysical and non-subjectivist explanation of emergent cultural practice' (Williams, 1973a, p. 12). Crucially, sources of such emergent cultural practice need not be assumed to be those based in a 'new class'. He thus proposed a general basis for their recognition :

> that no mode of production, and therefore no dominant society or order of society, and therefore no dominant culture, in reality exhausts human practice, human energy, human intention. (1973a, p. 12)

This essay was revised before it was reprinted in *Problems in Materialism and Culture*. In the interim Eagleton's Althusserian critique of Williams was published. Eagleton had welcomed what he saw as Williams's more recent rapprochement with 'Marxism', but criticized the 1973 essay for failing to fully escape from the alleged 'liberal humanism' and 'epistemological idealism' (a charge recomposed in later intellectual debates as 'humanist essentialism') of Williams's earlier work.[45] In an apparent act of clarification (or insistence), Williams revised the above passage for its 1980 republication thus:

> that no mode of production, and therefore no dominant society or order of society, and therefore no dominant culture, in reality exhausts *the full range of* human practice, human energy, human intention *(this range is not the inventory of some original 'human nature' but, on the contrary, is that extraordinary range of variations, both practised and imagined, of which human beings are and have shown themselves to be capable)*. (*PMC*, p. 43; italicization of revisions added)

This passage is an extraordinarily revealing indication of Williams's location of his 'insistence' within contemporary debates. It speaks directly to Norris's characterization of his 'leaving room for' the consciously purposive subject. However, as is very evident here, Williams also insists that this space-provision does not constitute an appeal to a 'humanist essentialism'.

Moreover, even the 1973 formulation anticipates Williams's position in *Marxism and Literature* concerning language and subjectivity. Had Williams been pressed on his assertions concerning the relationship between 'practical consciousness' and language – as he was on the unconscious in *Politics and Letters* – he may have provided a similar reformulation; that is, that 'a full range' of practical consciousness would necessarily include a hegemonically subordinate one. While this is implicit in Williams's assumption that incorporation is a constitutive practice within any hegemony, and indeed that hegemony 'saturates consciousness', he is insufficiently explicit on this point in his discussions of practical consciousness.

4.3 Williams versus Birmingham cultural studies?

> In recent Marxist work there has been a significant conflict... between those who, from their work on forms, have converted all social practices into forms (substituting epistemology for ontology: a position already reached within the later stages of the formalist tradition – Frye [1973], McLuhan [1967b]),[46] and those others, who, retaining an insistence on direct social practice, have to restate, often radically, positions on ideology and on cultural hegemony, but also, and more crucially, positions on creativity and its sources and formations, to which the formalist tradition has delivered an inescapable challenge but to which, also, it has contributed important and indispensable evidence. (1976b, p. 502)

Williams might here be referring to his own dispute with Eagleton's Althusserian critique, or perhaps the emerging differences between himself and the Birmingham CCCS. The latter were revealed more explicitly in his 'The Paths and Pitfalls of Ideology as an Ideology', a critique of the CCCS publication, *On Ideology*.[47] This will be discussed below but it is important to note that it is an early deployment of the critique of formalism in conjunction with Williams's sociology of cultural formations.[48]

While Althusser's influence was considerable at the CCCS, it was tempered initially not only by the CCCS's independent reading of Gramsci but also by Williams's *latest* work as well. The most notable example of such influence was the adoption of Williams's model of emergent, oppositional and alternative counter-hegemonic cultural forms within the overarching frame of the youth subcultures research programme.[49]

Stuart Hall had assumed effective directorship of the CCCS in 1968. His intellectual journey had had significant parallels with Williams's. They had similar backgrounds in literary studies and had worked together in the 'first' and 'second' 'New Lefts'.[50] However, Hall's focus appears always to have been directed more fully to popular culture, even before his appointment to the CCCS on its foundation in 1964.

Crucially, unlike Williams, Hall had rejected Goldmann's genetic structuralism, although for a different reason from Williams's reconstructive criticisms. For Hall, Goldmann's focus on coherent world views was not applicable to the 'critical ad hoc level at which ideologies are brought to bear on specific situations and organize the experience of particular groups and classes of men [*sic*]' (Hall, 1971, pp. 29–30). This focus on 'the ad hoc' – rather than the reflectively coherent – dimensions of popular belief, especially in relation to popular culture, was to prove a major influence on the Birmingham agenda.

Hall sought to link his own reading of Gramsci, Althusser and Althusser's collaborator, Nicos Poulantzas, with a *semiological* approach based mainly in the work of Roland Barthes. For Barthes, the Saussurean conception of the sign could be applied to non-linguistic phenomena such as photographs. In *Mythologies* he argued for a second order system of signification where such signs signified – primarily by connotation – 'myths'. As linguistic signifiers denote signifieds, so such signs signify myths. Myths dehistoricize and some myths – especially those associated with nationalist rhetoric – seek to naturalize political ideologies.[51] Barthes called the systems of 'cultural' familiarity on which such connotations relied *codes*.[52]

Like the Barthes of *Mythologies*, Hall located his semiological analyses primarily within an 'unmasking' conception of ideology critique where the principal function of the ideology is understood to be such 'naturalizing' legitimations of an existing order.[53]

Hall extended to news(paper) photographs Barthes's semiological work on the immanent *formal* delimitation of the possible ways in which photo-advertisements are interpreted by their viewers/readers.[54] In parallel with Williams's typology of the possible relationships between a hegemonic order and cultural forms (Table 3.1), Hall established a triple ideal-type typology of reception – consisting of dominant, negotiated and oppositional decodings.[55] He then developed and applied this model to television, first in his now famous paper, 'Encoding/Decoding in the Television Discourse', and later in a highly detailed formal analysis of the 'text' of an episode of the BBC's flagship current affairs programme, *Panorama*.[56] He also argued for a fourth code within which media texts were produced and 'structured in dominance', to produce a difficult to avoid 'preferred reading'. This was a professional code which, while relatively independent of the dominant code, still operated within its hegemony by means of such practices as the achievement of 'transparency of communication' and the overaccessing of élites in news-story production.[57] Formal analysis of programmes could reveal only this professional 'preferred encoding'.[58] However, it so also revealed the 'preferred reading' or dominant decoding.

It is not difficult to see why Williams would be hostile to this model. Where his own work on hegemony focussed on the need for the hegemonically dominant to incorporate independently produced cultural forms, Hall's offered little counter-hegemonic prospect beyond a resistantly consumptive decoding or, implicitly, the overturning of the entire 'dominant culture'. The professional code and its preferred encoding would have been problematic for Williams as their necessary location within the hegemonic limits of the dominant offered no possibility of immanent emancipatory ideology critique. Hall had allowed little, if any, theoretical space for oppositional or alternative *en*coding. Williams had a far more generous interpretation than Hall of practices within the 'hegemonic' professional code. Hall's 'Encoding/Decoding' paper was explicitly presented as a critique of naïvely liberal prospects of improved communication as a social panacea. In contrast, Williams acknowledged the gains for informed citizenship achieved by British current affairs television programmes – explicitly mentioning *Panorama* – in the same year that Hall published what Williams would have undoubtedly regarded as a formalist critique of the programme.[59]

This difference in assessment speaks to fundamental methodological and normative differences. For Hall, abandoning Goldmann had entailed the adoption of a structuralist rather than a genetic structuralist conception of homological correspondence. In exegetical elaboration of Lévi-Strauss's conception of homologous relations, and especially the anti-objectivist critique thereof by Bourdieu, Hall introduced 'articulation', the term which replaces 'homology' in his own practice, thus:

> Bourdieu wants to treat the problem in terms of the mutual articulation of two discontinuous fields. Symbolic relations are not disguised metaphors for class relations: but nor are they "merely signifying". It is *because* they do symbolic work of a certain kind, that they can function as the articulation of another field – the field of class relations: and hence also do the work of power and domination. (Hall, 1978a, p. 29)[60]

This is indeed broadly consistent with Bourdieu's argument as he elaborates it in 'Symbolic Power', a work known to the CCCS from 1977:

> We must remember that ideologies are always *doubly determined*, that they owe their most specific characteristics not only to the interests of the classes or class fractions they express... but also to the specific logic of the field of production... This provides us with a means of avoiding the brutal reduction of ideological products to the interests of the classes which they serve (this 'short-circuit' is common in Marxist critics) without succumbing to the idealist illusion which consists in treating ideological productions as self-sufficient, self-created totalities amenable to a pure and purely internal analysis (semiology).
>
> The properly ideological function of the field of ideological production is performed almost automatically on the basis of the structural homology between the field of ideological production and the field of class struggle. (Bourdieu, 1991a, p. 169)

But that 'almost' in 'almost automatically' is a summary account of a much more complex proposition. Bourdieu provides one of his 'friendliest' explanations of this hypothesis in a 1984 article:

> There is a political space, there is a religious space, etc.: I call each of these a *field*, that is, an autonomous universe, a kind of arena in which

> people play a game which has certain rules, rules which are different from those played in the adjacent space. The people who are involved in the game have, as such, specific interests, interests which are not defined by their mandators. The political space has a left and a right, it has its dominant and its dominated, the rich and the poor; and these two spaces correspond. There is a homology between them. This means that, *grosso modo*, the person who in this game occupies a position on the left, *a*, is related to the person occupying a position on the right, *b*, in the same way that the person occupying a position on the left *A* is related to the person occupying a position on the right *B* in the other game. When *a* wants to attack *b* to settle specific scores, he helps himself, but in helping himself he also helps *A*. This structural coincidence of the specific interests of the delegates and the interests of the mandators is the basis of the miracle of a sincere and successful ministry. The people who serve the interests of the mandators well are those who serve their own interests well by serving the others; it is to their advantage and it is important that it should be so for the system to work. (Bourdieu, 1991b, p. 215)

Thus rather than an interpretative Lévi-Straussian tabular construction of the binary oppositions between the 'mythemes' of a traditional myth, Bourdieu here accepts the institutionally given binary oppositions of objectified 'representative' conflict. These 'givens' enter the analysis as the first set of oppositions of the proposed homologous relation. Lévi-Strauss would diagrammatically represent Bourdieu's relations above thus:

a:b :: A:B

For Bourdieu, what prevents such a homological analysis falling into the 'brutal reductivism' of the Marxist 'short-circuit', or the idealist reductivism he attributes to semiology, is the granting of determinacy to 'the specific logic of the field of production' within the superstructures; that is, the intellectual composition of 'ideologies' and other cultural forms.[61] Hence the 'double determination' he proposes for ideologies.

Hall's 1976 analysis of the professional code of *Panorama* remains the most eloquent demonstration of his own 'homologous' analysis. Here two 'discontinuous fields', in Bourdieu's sense, were painstakingly analysed: the parliamentary theatre of party politics and the 'rules' of current affairs political reportage, and especially interviewing, as retrieved by semiological analysis.[62] Broadcast current affairs is shown not to be susceptible to conspiratorial charges of 'bias'. Rather, it is precisely its limited autonomy – including its norms of balance and objectivity – which demonstrates the homologous relation Hall proposes. This can be characterized by the following 'Lévi-Straussian' model:

State:political sphere :: political sphere:media

Thus

> Some such interpretation suggests that the relationship of the media to the political is remarkably *homologous* to the *general* relationship between politics and the State itself, in which politics (party practices) accords to the State (the institutions of power such as Parliament and the Courts) a certain measure of independence and neutrality, because this appearance is, ultimately, the most effective way in which politics can use or make itself effective *through* the State, without appearing directly to do so in the defence of narrow or short-term [c]lass or Party advantage . . . This is the sense in which both Gramsci and Poulantzas[63] speak of the State as necessarily a "relatively independent" structure. It is by the displacement of class power through the "neutral and independent" structures of the State, that the State comes to provide the critical function, for the dominant classes, of securing power and interest at the same time as it wins legitimacy and consent. It is, in Gramsci's terms, the "organizer of hegemony". If, then, we consider the media in homologous terms, we can see that they, too, do some service to maintenance of hegemony, precisely by providing a "relatively independent" and neutral sphere . . . And this reproduction is accomplished, not in spite of the rules of objectivity (i.e. by "covert or overt bias") but precisely by holding fast to the communicative forms of objectivity, neutrality, impartiality and balance. (Hall *et al.*, 1976, p. 88)

The gains here over conspiratorial formulations are considerable, but it is also the case that this model tends to place intellectuals (in Gramsci's expansive sense) in a remarkably instrumental role. If the homological correspondence between journalistic norms and the state is so neat, then journalists would seem to have little means of producing even the instabilities in the equilibria of hegemonic consent which, as Hall also acknowledges, Gramsci regards as inevitable. There is also simply no space in this account for the norm of informed citizenship that Williams invoked. This too would be seen as trapped within the hegemony of Hall's dominant code.[64]

In 1977 Hall published a remarkable essay, 'Culture, the Media and the "Ideological Effect"' (Hall, 1977b), which locates the above arguments within a broader account of the social production of hegemony. It is a very powerful summation of his own developing position and is the only 'Birmingham' work by him cited in the bibliographies of *The Sociology of Culture* and *Marxism and Literature*. There Hall makes his first characterization of Williams's 'definition of culture' as 'anthropological' and, ironically, *contrasts it* with the Marxian conception of productive force and, in a further irony, groups it with the anthropological conception of culture informed by Lévi-Strauss's structuralism.[65]

Hall's repositioning of Williams may explain the associations later made in Birmingham between Williams and 'ethnography' such as that within the research programme on youth subcultures. The dominant strand of this programme had explicitly linked Barthes's conception of code with Lévi-Strauss's *bricolage* (the process of 'primitive' mythmaking by reassemblage of available elements). The key methodological mechanism was the 'reading' of signifying elements of subcultural style such as modes of dress *as texts* within a subcultural 'repertoire' of 'negotiated solutions' to, for example, a contradictory class location. These actions were thus read as a *bricolage* of resistant significations, and characterized as one set of responses from within a subordinate culture resistant to a dominant (hegemonic) one.

But Lévi-Strauss's *bricoleurs* operated within an 'unconsciousness' which derives directly from the Saussurean distinction between langue and parole and, as Ricoeur has usefully pointed out, is 'more a Kantian than a Freudian unconscious' (Ricoeur, 1974, p. 33). That is, it resides within the deep structure that Lévi-Strauss tends to locate within a generic human mind. Crucially, the *bricoleurs* are in this sense not the authors of their actions.

Likewise, the subcultural participants were deemed to be similarly unaware of the full significance of such resistant practices. To couch the issue in a formulation derived from Althusser, they were living within 'imaginary relations' to their 'conditions of existence' (Clarke *et al.*, 1977, p. 48). Although Hall elsewhere separated Lévi-Strauss's position from the Althusserian conception of ideology, *Resistance Through Rituals: youth subcultures in post-war Britain* effectively brought the two together.[66]

When asked about the relevance of Birmingham's dominant/subordinate culture model to his own early work in adult education and 'the culture of the labour movement' in a 1977 interview, Williams himself raised the issue of the CCCS's subcultural research. While acknowledging the validity of researching such subcultures, he also reasserted the position first enunciated in his 'common culture' critique of Hoggart's conception of working-class culture, that the notion of a subordinate culture could not apply to intellectual traditions. Moreover, he stated, even if there had been spectacular subcultures in evidence in the 1950s, he would still have argued that 'the main battle is within the dominant culture' (1977a, p. 13).[67]

There is, of course, a sense in which Williams's criticism was at cross purposes with the CCCS research. For the Birmingham researchers, the application of the model of the code to working-class audience television reception and to subcultural practices was driven in part by a need to disprove dominant conceptions – largely associated with functionalist sociology – of a passive media audience, a politically quiescent working class and 'deviant youth'. Thus evidence of semiological 'resistance' demonstrated an active audience, a less than totally quiescent working class, and conduct by youth that was more socially significant than an easily dismissed 'delinquency'.[68] The Birmingham researchers were responding to socio-political developments similar to those Williams had characterized as the hegemonic incorporation of formerly oppositional cultural forms. They were also seeking to demonstrate the internally contradictory character of this incorporation, its 'unstable equilibria'.

But the author of 'Culture is Ordinary' did not see any need for disproof of the thesis that working-class people were rendered 'cultural dopes' (to use the negative phrase later popularized within cultural studies), even when working-class institutions moved from oppositional to incorporated positions. Such a commitment was an ethical given. Instead, Williams saw a danger that theoretical positions based in the 'alienated' structuralist assumptions detailed in the previous section could undermine the ethical basis of such commitments and replace it with *instrumentally* rational ones. Solidarity, Williams implied, should not require the same forms of empirical proof as 'objective' social science. Rather, it should become an informing component of a normatively critical project. This expectation becomes explicit in his critique of the CCCS.

Although Williams's 1977 review concerned the overtly theoretical CCCS text, *On Ideology*, he leaves the reader in no doubt that it is also the 'applied' CCCS work that he is challenging, since he prefaces his critique with a parody of a Barthes/Hall 'rhetoric of the image' analysis of the cover photograph. His key statement, laced with ironic references, is the following:

> The relative distance imposed by theoretical review permits the assimilation of selected evidence to the most diverse positions and procedures under the single title of "ideology", which seems to me now to mask rather than clarify the most urgent and most serious theoretical and practical choices...
>
> In [some] essays, there is an evident tension between empirical historical analyses of systems of ideas and their social sources and consequences and, on the other hand, models of ideology as coherent and totalizing (masking) systems, which can be discussed in terms of texts and codes, with metaphors such as "repertoire" from formal elements of play.
>
> In much recent theoretical work, including most of the essays in this volume, the latter mode is dominant and social relations and social movements tend to be seen through the procedures of a kind of textual analysis, for which an already-selected, and thus reciprocally-confirming, version of history is the (marginal) text.
>
> These relations and movements are "seen through", also, in another sense, since the dominant tone is dismissive, with a confidence that follows from the analytic (unmasking) procedures, rather than from any

> declared and substantial alternative position and policy. And this can be seen, in its turn, as the ideology of a group driven back, in an exceptionally frustrating period, from significant political intervention and participation, but regrouping within certain kinds of educational institution and intellectual work. (Williams, 1977b)

Williams's choice of the term 'unmasking' here coincidentally delimits unmasking ideology critique in much the same way as Márkus's contrasting of it with emancipatory ideology critique.[69] As we have seen, Williams only employs the latter 'in practice' but here it is heavily implied. The relative distance thesis is developed further in *The Sociology of Culture*. Williams is here, however, positing an analogical correspondence (Table 3.2) between the political alienation of the CCCS and the alienated assumptions of structural linguistics examined in the previous section. The hostile reference to 'texts and codes' echoes strongly a similar passage in *Marxism and Literature*.[70]

It has been argued that Williams confuses the semiological conception of code with that of 'encrypting'.[71] This is a plausible complaint given that, in *Keywords*, Williams even refers to the meaning of 'code' as 'opaque' (*KW2*, p. 307). However, in such references he is alluding to the use, as demonstrated by Hall, of Barthes's conception of code *in conjunction with* an Althusserian conception of ideology and/or Lévi-Straussian conception of unconsciousness that renders the code's 'subjects' blind to its rules. His only explicit reference to Barthes's conception of code is far more positive and stresses its potential to move beyond 'the closed categories of structural linguistics' (1976b, p. 504). This comment prefigures his later (and better known) one in a 1983 retirement lecture, that 'a fully historical semiotics would be very much the same thing as cultural materialism' (*WIS*, p. 210).[72]

The chief problem for the formational component of Williams's critique of the CCCS is his extension of the charge of alienation into an accusation of political disengagement. This cannot be reconciled with the publication the following year of *Policing the Crisis: mugging, the state and law 'n' order* (Hall *et al.*, 1978a). This sprawling work, already anticipated in CCCS working papers, was arguably the culmination of the CCCS media and subcultural research programmes. *Policing* also provided most of the conceptual framework for Hall's influential political writings of the 1980s.[73] It successfully predicted many features of 'Thatcherism' prior to Thatcher's election. One of its more contentious propositions was that an 'authoritarian consensus' was being developed in tandem with attempts to resolve a crisis of British hegemony by an expansion of the coercive role of the state. The identification and policing of 'mugging' was both a 'signifier of the crisis' and a key stage in the development of this authoritarian consensus. One could not find a more exemplary declaration of 'significant political intervention and participation' than *Policing The Crisis*'s prefatory expression of solidarity with the victims of the racism that was part of this law and order campaign.[74]

The conception of hegemony employed in *Policing* had been fully developed in Hall's 'Culture, the Media and the "Ideological Effect"'. The key step is the effective replacement of Williams's model of hegemonic incorporation by Poulantzas's understanding of the role of ideology in the hegemonic relation between the state and civil society. Hall implies that this was necessary because of Williams's 'continuing stress on experience and intention' (1977b, p. 332).[75]

Hall calls Poulantzas's approach 'separating and uniting' (1977b, p. 336).[76] By this he means, predominantly, a 'masking-fragmenting-uniting' process involving the fragmentation of 'classes into individuals' and the 'imposing of an imaginary unity or coherence' of a ruling ideology and the related field of political consensus. The media play a crucial role in 'winning consent' to these dominant ideologies. And the systematic 'penetration' and 'inflection' of the dominant ideologies into 'the discourses of the media' is achieved by the 'preferred codes' described above (Hall, 1977b, pp. 336–46).

Far from welcoming the increasing political engagement of the CCCS researchers, Williams's view of this project hardened during the 1980s. In 1984 he stated publicly, 'I don't agree with Stuart about authoritarian populism' (Williams, 1984).[77] This is consistent with the accusation in 'The Uses of Cultural Theory' concerning 'petty bourgeois intellectuals' who made 'long-term adjustments to short-term situations': 'that one form of theory of ideology produced that block diagnosis of Thatcherism that taught despair and political disarmament in a situation that was always more diverse, more volatile and more temporary'. (1986, p. 30; *POM*, p. 175)[78]

The publication of that lecture in *NLR* followed a significant debate in the journal in which Hall had answered similar criticisms of his authoritarian populism thesis.[79] Hall's adoption of 'authoritarian populism' in 1980 had marked a refinement of that thesis. As the post-war 'corporatist' consensus period of British politics ended with Thatcher's election, so Hall shifted his attention to the New Right's capacity to articulate its 'anti-statist' position with the receptive dimensions of working-class 'common sense'. Yet the model for this receptivity within 'authoritarian populism' remained much the same as it had in 'Encoding/Decoding'.

In 1988 Hall argued that the charge of 'pessimism' was understandable given the prime purpose of his work on Thatcherism (1988, p. 11). However, he embraced Gramsci's famous aphorism, 'pessimism of the intellect, optimism of the will', as a need for a new political will grounded in concrete analysis which so 'avoids the spurious oscillations between pessimism and optimism' (1988, p. 13). He argued that his acknowledgement of Thatcherism's 'hegemonic form of politics' was intended to awaken the need for a 'counter-hegemonic strategy' that recognized the changed terrain of political contestation (Hall, 1988, p. 11). This interestingly echoes Hall's acknowledgement that Orwell's *1984* should not be read too literally as it was intended 'less as a prophecy, more as a *warning*' (Hall, 1983b, p. 5). Williams, in contrast, came close to despising Orwell, and never accepted such a defence of either *1984* or *Animal Farm*, accusing him of 'passivity' and 'cutting out the spring of hope' (*O*, p. 78).[80] Indeed, the rhetorical pitch of Williams's criticism of (presumably) Hall in 'The Uses of Cultural Theory' is strikingly reminiscent of his attacks on Orwell.[81]

In both cases Williams's critical expectation is that their future projections – whether in politico-cultural analysis or fiction – should include plausible prospects for hope, whether this be found within the capacity for courageous action in projected contradictory moments, or within the 'actual' contradictory features of any given historical moment.

Williams's critique of Orwell is itself open to the legitimate criticism that it underestimates the formal properties of the dystopian mode, one of whose conventions is the hyperbolic representation of the societal flaws under criticism. Undoubtedly, any reply to this criticism from Williams would have employed his formational analysis. The issue for him is not the 'inherent' qualities of the form, but the choice of this form by an author who had other resources and options available. For Williams, his recognition of authorial agency entails his entitlement to stringent critique of the exercise of that agency.

Likewise, the formational analysis of the CCCS links the necessary social distance of academic – and especially theoretical – labour with the choice of an instrumentally calculative 'textualization' of social practices tied to an unmasking – albeit in many ways non-reductivist – conception of ideology. Stuart Hall has recently conceded that the central role of such a 'neutral' or instrumental conception of ideology within his work is closely related to the influence of the strategic pragmatism of Gramsci.[82]

This concession, I would suggest, enables a retrospective outline of the key differences between Williams and Hall. Each is committed to a form of 'prospective analysis' that builds from both *The Eighteenth Brumaire* and Gramsci. Each so recognizes the political significance of 'culture' as a terrain of hegemonic contestation. Each acknowledges that a complex non-reductivist analysis of this field is feasible and necessary. Each supports a political project characterized as counter-hegemonic. But, where Williams insists on the tentative identification of the democratizing 'resources of hope' within this counter-hegemony, Hall – at least within Williams's lifetime – prefers a more normatively neutral and strategic mode of writing that relies on a 'dystopian' invocation. But this key difference between the two was mutually misrecognized, usually as a narrowly conceived methodological one.

We saw this misrecognition played out in Hall's highly influential critiques of Williams discussed in Chapter 1. Against fairly obvious 'textual' evidence to the contrary, Hall insists that Williams abandons 'literary moral discourse' after *Culture and Society*, and thereafter confronts high culture by 'rendering culture ordinary' (Hall, 1997, p. 29). Hall, in effect, projects onto Williams's conception of culture the 'neutrality' of the conception of ideology

Hall adopted from Althusser (and/or Lévi-Strauss) and his own focus on popular culture.[83] But Williams maintains a privileged place for 'high' culture as autonomous art (and learning) precisely because of its relevance to a less instrumental conception of the counter-hegemonic. As he puts it in closing the chapter on hegemony in *Marxism and Literature*:

> cultural process must not be assumed to be merely adaptive, extensive, and incorporative. Authentic breaks within and beyond it, in specific social conditions which can vary from extreme isolation to pre-revolutionary breakdowns to actual revolutionary activity, have often in fact occurred. And we are better able to see this, alongside more general recognition of the insistent pressures and limits of the hegemonic, if we develop modes of analysis which instead of reducing works to finished products, and activities to fixed positions, are capable of discerning, in good faith, the finite but significant openness of many actual initiatives and contributions. The finite but significant openness of many works of art, as signifying forms making possible but also requiring persistent and variable signifying responses, is then especially relevant. (*M&L*, p. 114)

So the model of autonomous art demonstrates the contingency of hegemonic incorporation. Williams and Hall, thus, also fundamentally disagree about whether a practice – or an agent/subject – can be 'outside ideology'.[84] Williams's understanding of the relationship between culture, signification and hegemony effectively renders Hall's unmasking conception of ideology redundant. The elements of Williams's sociology of culture are informed by a need to further specify such contradictory and potentially counter-hegemonic situations. However, Williams seeks to identify all signifying practices which meet conditions of autonomy rather than merely re-employ orthodox 'identifications' of art. Thus Williams can normatively discriminate between, for example, autonomous and non-autonomous *popular* culture, as he demonstrated in his television criticism.[85]

4.4 Social formalism and cultural forms

One legacy of Birmingham cultural studies is an expanded conception of culture which takes its model of culture, in part, from the products of the culture industry. So the problem of autonomous cultural production is largely set aside for the varyingly politicized one of 'resistant reception'.[86]

Williams, in contrast, maintains a normative conception of cultural production as 'autonomous composition' (and, rarely, objectification), often at the expense of any comparable theoretical attention to popular reception. While he called for a 'rigorous sociological distinction of the variation and varieties of the popular' in 1976, popular reception – especially the limited ethnographic form at Birmingham – has a kind of 'absent presence' in

Williams's sociology of culture (1976b, p. 504). As we saw in Section 1.2, this lacuna developed quite early in his project.

Williams's declaration of commonality with a 'fully historical semiotics' would appear to leave open the prospect of a reconciliation with the Birmingham reception studies. But the compatible research programmes to which Williams alludes in *Marxism and Literature* suggest that a more likely fellow traveller would be a social formalist aesthetics that similarly under-values discrete reception studies.[87]

This emphasis is confirmed in Williams's construction of social formalism. To anticipate the argument below, Williams effectively replaces Hall's understanding of *code* – the likely basis, as we have seen, of Williams's charge of formalist conversion of 'all social practices into forms' in cultural studies – with a conception of *cultural form* developed from a sociology of genres.

Williams acknowledged as late as 1983 that (technicist) formalism was preferable to 'a Marxism which treats form as the "mere expression" or "outward show" of content' (*KW2*, p. 140). Likewise, *The Sociology of Culture* acknowledges that the Russian formalists justifiably defined their project against a 'sociological' approach because:

> What "sociological" then meant was either concentration on the general conditions of a practice, to the partial or total neglect of the practice itself; or, more immediately, appropriation of works in terms of their manifest or presumed social content, which was then assimilated to social content deduced from quite other sites, thus obliterating the most specific (and then it was said, the most formal) properties of the work or kind of work. (*SOC*, p. 139)

However, the formalism that resulted 'tended merely to reverse the priorities of its adversaries'. Social formalism is clearly intended to provide a solution to both sets of valid objections to the limited alternatives perceived by the Russian formalists. This section will focus on Williams's development of an alternative to (Russian) formalism and 'synchronic structuralism'.

Williams's characterization of the failings of formalism is broadly consistent with the charge Goldmann made against Lévi-Strauss's structuralism: 'a formalistic system that tends to eliminate in a radical way all interest in history and the problem of meaning' (Goldmann, 1973, p. 12).

Marxism and Literature is a key site for the development of the case for social formalism. With 'criticism' methodologically abandoned, Williams there examines the fate of 'literature' and 'aesthetic'. He later referred to these reflections as a 'clearing operation' of 'Cambridge' categories.[88] Williams wishes to reverse the contraction of literature to the 'specializing social and historical category' which had legitimated the understanding of (literary) criticism as discriminating judgement.[89] Rather, 'literature' should recover something like its 'pre-bourgeois categorization' as the 'range of actual writing' with a similar recovery of focus on composition. Here Williams echoes the critique of the artwork as object with which he closed the 'Base and Superstructure' essay.

Williams later admitted that the related chapter of *Marxism and Literature* entitled 'Aesthetic and other Situations' was the most difficult to write.[90] Once again Lukács is respectfully set aside. His conception of aesthetic specificity is deemed, like 'totality', to be too 'categorical' – that is, subject to *a priorism* – for the task Williams sets himself. That task is no less than the addressing of 'the multiple world of social and cultural process' within Williams's social theoretical ambitions. The (Russian) formalists' conception of 'literariness' based in literary 'devices' is seen to face a similar *a priorism*:[91]

> It is never the categorical distinction between aesthetic intentions, means, and effects and other intentions, means, and effects which presents difficulties. The problem is to sustain such a distinction through the inevitable extension to an indissoluble social material process: not only indissoluble in the social conditions of the making and reception of art, within a general social process from which these cannot be excised; but also indissoluble in the actual making and reception, which are connecting material processes within a social system of the use and transformation of material (including language) by material means. The formalists, seeking "specificity", in their detailed studies, not in a category but in what they claimed to show as a specific "poetic language", reached this crucial impasse earlier and more openly. One way out (or back) was the conversion of all social and cultural practice to "aesthetic" forms in this sense: a solution, or displacement, since widely evident in the "closed forms" of structuralist linguistics and in structuralist-semiotic literary and cultural studies. (*M&L*, p. 152)

This 'conversion' thesis is now familiar. Its reverse form – the turning of aesthetico-critical theses into de facto social theories – is usually labelled 'projection' by Williams.[92] It is in avoidance of such projection that Williams turns instead to the 'more interesting way out' offered by the Prague structuralists: 'to move the definition of art to a "function" and therefore a "practice"' (*M&L*, p. 152).

Jakobson had developed a model of 'the dominant' amongst functions within language.[93] The literary is the least instrumental of these functions since it focuses on the formal means of communication 'for its own sake'. Literature was thus defined as those uses of language in which the literary function was dominant. But it was Mukařovský who extended this model to a more generalized 'aesthetic function'.[94]

Williams embraces Mukařovský's 1936 work, *Aesthetic Function, Norm and Value as Social Facts*.[95] There Mukařovský advances the proposition that the aesthetic function is widely distributed socially but may not always be the dominant function. Art is defined by the dominance of the aesthetic function. He insists that the aesthetic function is not an inherent property of an object but a product of a 'collective awareness', systems of norms (including aesthetic norms) that reveal themselves 'by exerting a normative influence on empirical reality' (Mukařovský, 1979, p. 20). So Mukařovský provided Williams with both a confirmation of his own break with the existent standards of 'criticism' and, it would seem, a theoretical basis for his replacement of the categories of aesthetic text, work and object with 'practice' and 'form'. As we shall see in Section 5.2, he also provided him with a means of finally jettisoning the baggage of 'culture as whole way of life'.

Although Williams does not comment directly on this, it is also significant that Mukařovský criticized the Russian formalists for failing to recognize a 'thematic' semantic dimension as well as the aesthetic devices which constitute art. In 1934 he had rendered this argument semiotically by arguing that while art was an 'autonomous sign' it was also an 'informational sign'.[96] Mukařovský and Goldmann thus have some compatibility.[97]

Williams was also hostile to the Russian formalist conception of literary *system*. For him, synchronic structuralism's problems derived directly from the Russian formalist legacy of the conception of literary system modelled on Saussure's *langue*. Williams acknowledged Jakobson's and Tynjanov's attempted correction of this tendency. In this short text they usually use 'system' to refer to both linguistic and literary systems.

> The opposition between synchrony and diachrony was an opposition between the concept of system and the concept of evolution; thus it loses its importance in principle as soon as we recognize that every system necessarily exists as an evolution, whereas, on the other hand, evolution is inescapably of a systemic nature. (Jakobson and Tynjanov, 1971, p. 80)[98]

But Williams found this appeal to a conception of history as 'evolution' still trapped within the 'familiar reification of objective idealism' which needed amendment 'by the full emphasis of the social process' (*M&L*, p. 42).

Indeed, when Williams revisits Goldmann's work in a 1978 review he states explicitly for the first time that the continuing appeal of Goldmann's genetic structuralism lay in its capacity to deliver what Russian formalism could not:

> [genetic structuralism] shares with other positions an emphasis on *forms* as the centre of interest in cultural creation. The terms *form* and *structure* are indeed often interchangeable. But the position is distinct from formalism in that what is in question is always "the form of the content"... (Williams, 1978d, p. 26)

Williams's account of Goldmann here echoes Lévi-Strauss's famous self-defence of his analysis of myth against the charge of formalism (as represented by Vladimir Propp's analysis of folktales[99]):

> Contrary to formalism, structuralism refuses to set the concrete against the abstract and to recognize a privileged value in the latter. *Form* is defined by opposition to material other than itself. But *structure* has no distinct content; it is content itself, apprehended in a logical organization conceived as property of the real. (Lévi-Strauss, 1976, p. 115)

This partial convergence is interesting. Neither Lévi-Strauss nor Goldmann – at least as Williams reads him – conceive form/structure as ontologically distinct from semantic 'content'. Neither regards form/structure as a set of 'devices'. Of course the differences begin once the immanent structuring of 'content' is discussed. In the above, the 'logical organization' of content is, of course, that revealed by Lévi-Strauss's paired binary oppositions of 'mythemes'.

In a critique of and debate with Lévi-Strauss, Paul Ricoeur argued that Lévi-Strauss's method – and structuralism's synchronic bias generally – rendered impossible a role for a hermeneutics of 'content'.[100] Lévi-Strauss responded that it was not a question of exclusion but rather that 'the recovery of meaning is secondary and derivative compared with the essential work which consists in taking apart the mechanism of objectified thought', and that in any case, 'we cannot understand on the inside unless we were born on the inside' (Lévi-Strauss, 1970, p. 66). Ricoeur's critique was designed to place limits on the application of Lévi-Strauss's conception of myth 'unconsciously' created by *bricoleurs*. He argued that the structuralist analysis of totemic myths – where 'synchrony takes the lead over diachrony' – was valid but too much was lost semantically if the model of myth was applied to cultures with an active practice of interpretation and tradition. Hence he implied a division of labour between hermeneutics and structuralism. Lévi-Strauss refused 'the bargain' largely because it appeared to rest on a distinction between traditional and modern 'civilizations' that could not accommodate the continuities he saw between the two. In particular, he reformulated the problem thus:

> Are we dealing with an intrinsic difference between two kinds of civilization, or simply with the relative position of the observer, who cannot adopt the same perspective *vis-à-vis* his own civilization as would seem normal to him *vis-à-vis* a different civilization? (Lévi-Strauss, 1970, p. 61)

This dispute – and especially this passage by Lévi-Strauss – was picked up the following year by Gérard Genette and applied to the field of literature.[101] Genette stressed that the relationship between structuralism and hermeneutics might still be one of complementarity based in Lévi-Strauss's emphasis on observational positions. Genette's recognized that a structuralist literary criticism needed to legitimately address semantic phenomena and should not be 'confined to counting feet and observing the repetition of phonemes' (Genette, 1982a, p. 10).

This point of contention can be sharpened by employing Pettit's critique of the dependence of structuralism and formalism on the structural linguistic model of the sentence in their approach to narrative, most obviously exemplified by Propp's analysis as it was taken up by the early Roland Barthes.[102] Lévi-Strauss, as we can see in the passage above, rejects Propp for his phonologically derived model of binarisms. Genette acknowledges the sentential basis of Propp's formalism but is clearly dissatisfied with 'the model of the sentence' as an adequate account of literary semantics within a 'structuralist poetics'.

Genette is thus acutely aware of the risks as well as the gains of formalism and synchronic structuralism.[103] His reflections so provide a kind of 'missing link' for Williams's own deliberations on these questions. For not only do Genette and Williams exercise similar caution towards synchronicism, they also share an interest in the 'literary' genre as an alternative to the model of the sentence for the analysis of narrative forms.[104] Genette's solution is to locate genres as an 'anthropological' dimension of modern societies in a specific sense: namely, the recognition – that first emerged in classical poetics – of the relationship between genres and 'public expectations'. This too, as we shall see, he shares with Williams.

Genette's influential 1979 work, *The Architext*, critically rescues Aristotelian poetics from the goals attributed to it by neo-classicists and especially by the Romantics.[105] Genette's chief complaint about these genre theories is broadly the same as Márkus's of Marxian ones – their tendency to ground typologies of genres in 'natural' transhistorical forms.[106] Genette traces this tendency to a misattribution to Aristotle of a three-genre model of narrative, drama and lyric. He demonstrates that Aristotle instead recognized only two contemporary *modes*, narrative and drama. Modes are solely characterized by what Genette calls their 'situations of enunciating': only the poet speaks in the narrative mode while in the dramatic mode only the characters speak.[107] It is principally by the combination of modal choice with a highly hierachicized set of thematic 'objects' – 'content' – that 'the Aristotelian genre system' is constituted.

Accordingly, the tendency towards naturalization of 'fundamental' genres is, for Genette, a product of the Romantics' failure to distinguish genres from modes. Thematic and enunciative criteria are both 'naturalized' and hierarchicized. Modes in contrast are not 'properly literary' categories but, for Genette, *linguistic* ones. They are thus 'natural forms' in the sense of 'natural languages', that is, 'only to the extent that language and its use appear as facts of nature vis-à-vis the conscious and deliberate elaboration of

aesthetic forms' (1992, p. 64). Nonetheless, Genette is prepared to concede that 'a certain number of thematic, modal and formal determinations... are *relatively constant and transhistorical*' (1992, p. 78). It was these that he characterized in his earlier work as fundamental.

Genette's clarifications highlight the key difference between a genre analysis and a 'structuralist analysis of narrative': the restoration of diachrony to the synchronic bias of a 'sentential' analysis. The relation between a genre's 'history' and contemporary composition is more than the product of the intersection of syntagmatic 'choices' from a paradigmatic axis, as the sentential model necessitates.

It is in a similar context that Williams valued genetic structuralism's potential to encompass 'the full emphasis of the social process'. For Williams then, the tasks of social formalism include a recognition of the 'transhistorical' dimensions of genres. He so supplants the role of the formalist/structuralist self-reproducing synchronic system by, crucially, the provision of a means of socio-historically accounting for the generic innovations of 'devices', and more major formal/conventional innovations. In terms of the production paradigm (especially for Adorno), this constitutes a recognition of the *cumulative* determinacy of cultural forms as a *productive force*.[108]

Yet this recognition in turn raises in a different manner those complexities of social and historical correspondence discussed in previous chapters. Williams's key step is to provide his own typology of cultural forms specifically designed to recognize *differing* modes of correspondence of conjunctural cultural forms, including that determinacy exercised by 'trans-epochal' cultural forms.

As noted in Section 2.5, Williams too works with a conception of mode, but this is not derived directly from any Aristotelian schema. Williams would thus appear to escape Genette's major criticism of most other theorists who conflate the concepts of mode and genre. Williams's reassessment of genre theory had begun in the 'Base and Superstructure' essay. There he made it plain that analysis of cultural forms was crucial to his adoption of the notion of 'practice' instead of 'art object', but that the genre analysis of 'orthodox criticism' was inadequate, and that there need be no coincidence of 'collective modes' and genres.[109] *Marxism and Literature* provided Williams's own criticism of neo-classical and Romantic genre theories.[110] By *The Sociology of Culture* this position had gelled into the development of the typology in Table 4.1.[111]

For Williams, as we have seen, Goldmann's genetic structuralism was flawed by its confinement of correspondence to an 'epochal' frame. Williams's delineation of the varying levels of correspondence and conditions of reproducibility of the different subtypes of cultural forms is clearly designed to prevent the collapse of such analysis into 'epochal' reductivism. Moving down column 1, we can see that each sub-type brings us closer to specificity – not of 'the present' but rather to an abstract-historical specificity – by introducing

Table 4.1 Williams's typology of cultural forms

Form	Immanent formal properties	Correspondence (if any)	Examples
Mode	'Properly collective forms' constituted by highly complex external and internal *signals* capable of elaboration into contemporary genres	Relatively independent of specific social orders	Drama Lyric Narrative utopia[a] 'New mode' of cinema
Genre (Kind)	Specific activations and elaborations of 'modes' within definite social orders	Have *some definite dependence* on changes in epochal orders	Tragedy Comedy Epic Romance 'Fiction'
Type	Radical distributions, redistributions and innovations operating over relatively long periods within an epoch	*Correspondence* with 'specific and changed social character of an epoch'	'Bourgeois' drama Realist novel Landscape painting
Form	Radical distributions, redistributions and innovations linked to the smaller-scale social contradictions within an epoch; often tied to alterations of the typical	Subject to forms of organization of cultural producers/formation, cultural institutions and level of development of means of cultural production; major formal *innovations* have a further set of determinants	Breaks to naturalist drama and subjective expressionism in television; Soliloquy

[a] Williams (1978g). Cf. discussion in Chapter 7.

more and more external-social and *internal-cultural-formal* determinants. In effect, Williams has added what he sees as further necessary levels of mediation. Only with the confusingly named 'form' do we reach a 'sub-epochal' arena of formal change.

With concrete formal analysis, however, Williams is otherwise quite close to Goldmann (at his best). As we saw in Section 3.2, Goldmann's analysis of Racine's tragedies in *The Hidden God*, for example, did not attempt to establish an homology between world views and the 'content' of artworks understood as anything as crude as expressed politico-social statements. More specifically, the formal component of his analysis sought homologous relations between certain events, and changes in Racine's usage of formal 'devices' from play to play, culminating in the role of two Aristotelian features

of tragedy, 'peripeteia' (unexpected reversal of action) and 'recognition' in Racine's *Phèdre*.[112] As we saw in Section 3.4, Williams's critiques of literary forms placed a similar emphasis on formal devices understood, however, as *conventions*.

As can also be seen in Table 4.1, Williams activates his appropriation of the Vološinovian conception of signal (as distinct from sign) at the modal level, as 'highly developed and complex internal signals' (*SOC*, p. 194). In effect, the formalist conception of device is thus confined to that level, while at all others it is replaced by socially determinate 'internal' conventions (understood at their formation as formal innovations). Crucially, it is this that marks Williams's social formalism as 'cultural materialist' as opposed to Genette's structuralist delineation of the mode as a linguistic phenomenon. Indeed, Williams's typology of means of cultural production makes it evident that the 'external signals' of modes require elementary (non-linguistic) means of cultural production.[113] It is by this means too that Williams can 'modernize' genre theory's reach beyond a narrowly literary set of practices.

Williams's emphasis on the social production of forms nonetheless shares much with Goldmann's genetic structuralism. He implied this commonality in the 1978 assessment:

> This position is then distinguished from structuralism: first by an insistence on the presence of such *subjects* (the forms have this precise human and social embodiment); by a consequent insistence on the *functions* of such forms, in the actual relations and struggles of the groups which create them; and by a final emphasis on history, in which the "genesis" of such forms – their formation, maintenance and breakdown – is a central element of the changing totality of social life. (Williams, 1978d, p. 26)

The 'genesis', crucially, is not simply a matter of external social forces acting on a separate and discrete cultural form. Rather, the cultural form – or its formal innovation at key historical moments for Williams – is an emergent 'way of seeing' changed social relations.

There is no question that the paradigmatic case for Williams is drama which, as he makes it perfectly plain, is partly a product of his own research interest but also, fortuitously, one of the best-documented cases available of trans-epochal endurance of a (modal) cultural form. The entire chapter on forms in *The Sociology of Culture* is devoted to theses about the development of dramatic forms that Williams had previously advanced in other publications.[114] Two case studies he provides in elaboration of the above are useful here.

The first is a case study of the soliloquy. Williams traces the emergence and acceptance by audiences of the convention that a single speaker would reflect aloud in 'their' presence. He draws this methodological conclusion:

> These new and subtle modes and relationships were in themselves developments in social practice, and are fundamentally connected with the discovery, *in dramatic form*, of new and altered social relationships, perceptions of self and others, complex alternatives of private and public thought. It is true then that what has been discovered, and can later be analysed, in the form can be shown to be relatively associated with a much wider area of social practice and social change. New conceptions of the autonomous or relatively autonomous individual, new senses of the tensions between such an individual and an assigned or expected social role, evident in other kinds of contemporary discourse but evident also in analytic history of the major social changes of this precise period, are then in clear relation with the 'device'.
>
> But it is not necessary to explain the device as their consequence, taking first the sociology and then the form. This may often appear to be the order of events, but it is often also clear that the formal innovation is a true and integral element of the changes themselves: an articulation, by technical discovery, of changes in consciousness which are themselves forms of consciousness of change. Thus to analyse the soliloquy in English Renaissance drama is necessarily, first, a matter of formal analysis, but not as a way of denying or making irrelevant a social analysis; rather as a new and technically rigorous kind of social analysis of *this* social practice.
>
> We can then see the point at which formal analysis necessarily challenges previously limited or displaced kinds of social analysis. For while social analysis is confined to the society which, as it were, *already* exists, in completed ways, before the cultural practice begins, it is not only that analyses made elsewhere are simply applied to actual works, imposing on them only the most general considerations and missing or neglecting other elements of their composition. It is also that actual evidence of the general socio-cultural process, in one of its significant practices, is not even looked for, though it is in fact abundant. This is the point of transition for a sociology of culture, to include, as a major emphasis, the *sociology of forms*. (*SOC*, pp. 142–3)

The reference to 'displaced' social analysis clarifies the 'limit' Williams would place on his endorsement of that technique in *Marxism and Literature*.[115] When he returns to the case of the soliloquy later in the same chapter, Williams casually suggests a formal-comparative exercise of drawing up two columns: one listing formal characteristics of the soliloquy, the other listing 'general social changes in self-conceptions of the individual and in relations between individuals in this new sense and their assigned or expected social

Dates of First Publication and/or First Editions of Key Works by Williams

1950 *Reading and Criticism*
1952 *Drama From Ibsen to Eliot*
1953 'The Idea of Culture'
1954 *Drama in Performance*
Preface to Film (with Michael Orrom)
1956 'T.S. Eliot on Culture'
1957 'Fiction and the Writing Public' (critique of Hoggart)
1958 *Culture and Society*
'Culture is Ordinary'
1961 *The Long Revolution*
1962 *Communications*
1966 *Modern Tragedy*
1968 *The May Day Manifesto (ed.)*
'Culture and Revolution: a comment'
1969 'On Reading Marcuse'
1970 *The English Novel: from Dickens to Lawrence*
1971 *Orwell*
'Literature and Sociology: in memory of Lucien Goldmann'
1973 *The Country and the City*
'Base and Superstructure in Marxist Cultural Theory'
'Baudelaire's Paris' (review of Benjamin)
1974 *Television: technology and cultural form*
'On High and Popular Culture'
'The Frankfurt School' (review)
1975 'Drama in a Dramatised Society: an inaugural lecture'
1976 *Keywords*
'Developments in the Sociology of Culture'
'Notes on Marxism in Britain Since 1945' (cultural materialist 'manifesto')
'How Can We Sell the Protestant Ethic at a Psychedelic Bazaar?' (review essay on Daniel Bell)
'Communications as Cultural Science'
1977 *Marxism and Literature*
'The Paths and Pitfalls of Ideology as an Ideology' (critique of CCCS)
1978 'The Significance of "Bloomsbury" as a Social and Cultural Group'
'Means of Communication as Means of Production'
'Utopia and Science Fiction'

R&C	*Reading and Criticism*	(1950) London: Frederick Muller
ROH	*Resources of Hope*	(1989) Ed. R. Gable. London: Verso
RWOT	*Raymond Williams on Television: Selected Writings*	(1989) Ed. A. O'Connor. London: Routledge
SOC	*The Sociology of Culture*	(1995) Chicago: University of Chicago Press/Shocken Books. First published as *Culture*. London: Fontana, 1981
TEN	*The English Novel: From Dickens to Lawrence*	(1974) London: Paladin. First published by Chatto & Windus, 1970
T2000	*Towards 2000*	(1983) London: Chatto & Windus/Hogarth
TV1	*Television: Technology and Cultural Form*	(1974) London: Fontana
TV2	*Television: Technology and Cultural Form* (2nd Edn [ed.] Ederyn Williams)	(1990) London: Routledge
WICTS	*What I Came To Say*	(1989) London: Hutchinson Radius
WIS	*Writing in Society*	(1984) London: Verso